Lecture Notes in Electrical Engineering

Volume 2064

Mikhail V. Nesterenko • Victor A. Katrich • Yuriy M. Penkin • Victor M. Dakhov • Sergey L. Berdnik

Thin Impedance Vibrators

Theory and Applications

Springer

Mikhail V. Nesterenko
V.N. Karazin Kharkov National University
Dept. Radiophysics
Svobody Sq. 4
61077 Kharkov
Ukraine
Mikhail.V.Nesterenko@gmail.com

Victor A. Katrich
V.N. Karazin Kharkov National University
Dept. Radiophysics
Svobody Sq. 4
61077 Kharkov
Ukraine
vkatrich@univer.kharkov.ua

Yuriy M. Penkin
National Pharmaceutical University
Dept. Information Technology ·
Pushkinskaya st. 53
61002 Kharkov
Ukraine
penkin.yuriy@gmail.com

Victor M. Dakhov
V.N. Karazin Kharkov National University
Dept. Radiophysics
Svobody Sq. 4
61077 Kharkov
Ukraine
dakhovvm@gmail.com

Sergey L. Berdnik
V.N. Karazin Kharkov National University
Dept. Radiophysics
Svobody Sq. 4
61077 Kharkov
Ukraine
Sergey.L.Berdnik@univer.kharkov.ua

ISSN 1876-1100 ISSN 1876-1119 (eBook)
ISBN 978-1-4614-2799-5 ISBN 978-1-4419-7850-9 (eBook)
DOI 10.1007/978-1-4419-7850-9
Springer New York Dordrecht Heidelberg London

Printed on acid-free paper

Springer is part of Springer Science+Business Media (www.springer.com)

Even to deliver and explain what I bring forward is no easy matter; for things in themselves new will yet be apprehended with reference to what is old.

Francis Bacon

Preface

This book is devoted to the theory of thin vibrator antennas using an impedance approach, developed by the authors as a generalization of the classic theory of perfectly conducting thin vibrators. The notion of impedance, due to the wide spectrum of electrodynamics problems solvable by such an approach, has become one of the most universal methods for modeling wave processes. It provides a wonderful opportunity for analytical solution searching, simplifying the mathematical formulation of the boundary problem. Of course, the restrictions imposed on parameters during problem formulation and solution somewhat limit such a possibility. However, from a practical point of view, the necessary restrictions turn out to be natural for thin vibrator antennas. The universality of the impedance model and the naturalness of the antecedent suppositions have made it possible to create a theory of thin impedance vibrator antennas in a single framework as a generalization and continuation of classical electrodynamic theory for thin perfectly conducting vibrators. The scope of this book includes conducting cylindrical radiators, corrugated and ribbed vibrators, radiators with isolating coverings, and so on, and it covers different cases both of impedance vibrator antenna excitation and their location in spatial regions of different geometries, including regions filled by a material medium. This concept allows one to control and optimize the electrodynamic characteristics of vibrator antennas and finally makes it possible to widen the boundaries of the application of vibrator antennas application to complex, modern radio and electronic systems and devices, taking into account requirements concerning their mobility and reliability in a hostile environment.

It should be noted that the mentioned theory of thin vibrator antennas with distributed surface impedance has been formulated in separate papers (mostly by the authors of this monograph) and published over last 30 years in a variety of scientific publications. However, these results have never been generalized and assembled in a separate publication, and that impelled us to write the book.

This monograph comprises seven chapters and five appendices. References are given at the end of each chapter and are indicated by a number in square brackets, e.g., [3] in Chap. 2.

Chapter 1 is introductory. It briefly presents equations of macroscopic electrodynamics and the approximating analytical methods of their solution. We hope that this information will make it possible to read the succeeding chapters without consulting the specialized literature.

Chapter 2 is essentially methodological and is devoted to the key problem concerning radiation of electromagnetic waves by a thin impedance vibrator located in free space or in an infinite absorbing material medium. The asymptotic solution for the current on the impedance vibrator is obtained by the averaging method. The agreement of this solution with results obtained by other methods for symmetric perfectly conducting vibrators is analyzed in detail. Section 2.2.2 is worthy of special attention. Here asymptotic formulas for complex surface impedances representing different metal-dielectric structures are given for reference. They include a solid cylindrical conductor with skin effect, a periodically corrugated or ribbed cylindrical conductor, a cylindrical conductor with isolating magnetodielectric covering, a cylindrical conductor with transverse periodic dielectric insertions, and a dielectric cylinder. These formulas relate vibrator structure and the numerical value of the surface impedance, which may be used as a parameter in a mathematical model.

Resonant properties of impedance vibrators in dependence on surface impedance value are also analyzed in this chapter. It should be noted that the formula for resonant vibrator length allows for the evaluation of its value by surface impedance only for the required degree of radiator miniaturization. Radiation fields of resonant impedance vibrators in infinite medium are also covered in this chapter. Since impedance vibrators are widely used in medicine, the near-field analysis is made for biological tissues.

At present, impedance vibrator antennas, particularly in multilayered dielectric shells, are used in various medium with electrophysical parameters sufficiently different from those of air. The vibrators are applied in devices for underground and undersea radio communication, medical diagnostics and hyperthermia, geophysical investigations, and so on, and are located near metallic bodies of various configurations. The simplest configuration consists of an infinite perfectly conducting plane, which, in turn, is a good model for the approximation of screens in the form of finite-dimensional plates.

In Chap. 3, we consider the problem of radiation of electromagnetic waves by impedance vibrators in material medium located over an infinite perfectly conducting plane. Since the problem for a vertical asymmetric vibrator (monopole) reduces by a mirror-image technique to that for a symmetric vibrator, as analyzed in Chap. 2, a horizontal vibrator is chosen for this study. In this chapter, we obtain by the averaging method approximate analytical formulas for currents and full electromagnetic fields exited by a single horizontal impedance vibrator and by a system of crossed vibrators over a perfectly conducting plane in a half-infinite homogeneous medium with losses. The spatial distribution of the near fields in dependence on the medium's material parameters and the influence of the plane is investigated numerically. The formation of a radiation field with given spatial-polarization characteristics by a system of crossed impedance vibrators is also analyzed.

The vibrator's cross section and the distribution of surface impedance can serve as additional parameters to obtain the required electrodynamic characteristics for cylindrical vibrator antennas. Such rather specific, at first glance, nonregular vibrators are treated in Chap. 4. Here, to avoid overlapping with results in the literature while retaining the generality of the problem, the problem of excitation of a nonregular vibrator in free space by an incident-plane electromagnetic wave is considered. The normalized backscattering cross section is chosen as a parameter in a numerical simulation. In addition, this chapter deals with some other questions as well. As expected, structural complications of vibrators lead to difficulties in obtaining approximate analytical solutions for the vibrator current. These solutions, except in some specific cases, become too cumbersome and of little use in engineering practice. Such a situation arises, for example, in the analysis of multielement vibrator antennas. Therefore, in Chap. 4 a new numerical-analytic method for solving scattering problems, the generalized method of induced electromotive forces (EMF), is proposed and validated. The novelty of the proposed method lies in the fact that the functional dependencies found by analytical solutions, preliminarily obtained for the vibrator current by the averaging method, are used as the basis function (functions) for approximation of the vibrator current.

In Chap. 5, we investigate, by the generalized method of induced EMF, several multifaceted problems (for an impedance vibrator with arbitrary excitation point in free space; for an impedance monopole in a material medium; for a vibrator with symmetric and antisymmetric components of impedance along its length; for a system of impedance vibrators), thus proving the efficiency of our proposed method for solving various problems in electrodynamics. Numerical results are compared with the solutions, and experimental data obtained by other investigators and comparison results are thoroughly analyzed.

Thin impedance vibrator antennas are widely used in various mobile objects, including air and space vehicles. The bodies of these mobile objects often have a spherical form, or they can be approximated by a sphere. Here the dominant configuration is an asymmetric radially oriented monopole. Naturally, a mathematical model of such antennas in a strict electrodynamic formulation for a particular vibrator geometry and a spherical object has a large practical significance. Therefore, Chap. 6 presents a problem on the radiation of electromagnetic waves by a thin impedance radial monopole located on a perfectly conducting sphere. The solution here is derived by approximate analytical expression for the monopole current in terms of the spherical Bessel functions, allowing, with the Green's function for the space outside the sphere, integration of the expressions for the vibrator fields in spherical coordinates by known formulas. Of course, an approximation is also essential for application of the generalized method of induced EMF to vibrator systems located on a sphere. In this section, we also study impedance vibrators, both for arbitrary location of the excitation point on the vibrator and for a monopole fed at the contact point of a sphere and a monopole. The radiation fields for the vibrator antenna are also represented here.

In contrast to previous chapters, where outer excitation problems for impedance vibrators are studied, Chap. 7 deals with inner electrodynamics problems for vibrator structures located in a rectangular waveguide. The aim here is twofold. Firstly, we would like to demonstrate how the methods proposed in this monograph are applied to problems in closed spatial regions, namely in a rectangular waveguide. The choice is related to the fact that vibrator elements are widely used in waveguide devices in different applications, and the problems have independent applied significance. Secondly, we want to confirm new theoretic results for impedance vibrators by original experimental investigations, which in laboratory environments prove to be simpler and more reliable than waveguide experiments. It should be noted that these aims have been successfully attained by investigating electromagnetic wave scattering by vibrators with constant and variable surface impedances and also by impedance vibrators of variable radius.

Appendix A contains the formulas for the electrical Green's functions for various electrodynamic volumes considered in this monograph. In Appendix B, the basics of the method of moments extensively used for the numerical solution of the integral equations for currents on vibrator radiators are briefly outlined. In Appendix C, the formulas for generalized integral functions are given, and in Appendix D, we describe a method of series summation for functions of the self-field of a vibrator in a rectangular waveguide. Since the absolute Gaussian system of units (CGS) is used in this monograph, Appendix E gives conversion factors for equations and units between the CGS and SI systems.

The authors refuse to finish this book with the traditional terminal punctuation sign, the period or "full stop". Instead, we close with a more-optimistic ellipsis. In this final section we propose a list of problems of "nearest perspective" that can be solved for thin vibrators in the framework of the theoretical approaches presented in the book.

This monograph is written in an academic style, and it can be used by students and graduates of technical universities. It should be noted that any of the current applied problems presented in the concluding section may serve as the basis of a doctoral dissertation. On the other hand, the results of numerical modeling and asymptotic formulas obtained by the authors will be useful for practicing engineers and designers of antenna and waveguide systems, including those for mobile objects and medical devices operating under conditions of a demanding environment.

The authors consider it their pleasant duty to express their gratitude to Nadezhda N. Dyomina and Anatoliy M. Naboka for editing the English text.

Kharkov, Ukraine

Mikhail V. Nesterenko
Victor A. Katrich
Yuriy M. Penkin
Victor M. Dakhov
Sergey L. Berdnik

Contents

Chapter 1
General Questions of the Theory of Impedance Vibrators in the Spatial-Frequency Representation

In this chapter, the main equations of macroscopic electrodynamics, the theory of thin impedance vibrators and the approximate analytical methods of their solution, and expressions for the tensor Green's functions of various spatial regions are briefly presented. The materials presented here will be used throughout the book, allowing readers to use the book without any additional references.

1.1 Problem Formulation and Initial Integral Equations

Let us formulate a general problem of electromagnetic wave scattering (radiation) by a material body of finite dimensions. The problem geometry and corresponding notations are represented in Fig. 1.1. Let an electromagnetic field be generated by extraneous sources $\{\vec{E}_0(\vec{r}), \vec{H}_0(\vec{r})\}$, where $\vec{r}$ is the radius vector of the observation point, is incident on a body with homogeneous material parameters (permittivity ε, permeability μ, and conductivity σ), occupying a volume V, and bounded by a smooth closed surface S. The time-dependence for the entire field is represented as $e^{i\omega t}$ ($\omega = 2\pi f$ is the circular frequency; f is the frequency in Hertz). The body is located inside the electrodynamic volume V_1, bounded by a perfectly conducting (or impedance) surface S_1, which may be infinitely distant. The volume V_1 is filled by the medium with parameters ε_1, μ_1 expressed in the general case as complex piecewise-constant functions depending on the coordinates. The field of extraneous sources may be given as an electromagnetic wave field incident on the body (scattering problem) or as a field of electromotive forces (EMF) applied to the body, nonzero only in a portion of the volume V (radiation problem), or in the general case as a combination of these fields. It is necessary to find the full electromagnetic field $\{\vec{E}(\vec{r}), \vec{H}(\vec{r})\}$ in the volume V_1 satisfying Maxwell's equations and boundary conditions on the surfaces S and S_1.

It is well known that the problem thus formulated can be treated by solving equations for the electromagnetic fields in differential or integral form. The fields derived by the solution of integral equations are known automatically to satisfy boundary conditions on the body's surface [1]. Moreover, they are very effective if the boundary surfaces S and S_1 are coordinate surfaces in different coordinate

M.V. Nesterenko et al., *Thin Impedance Vibrators*, Lecture Notes in Electrical Engineering 2064, DOI 10.1007/978-1-4419-7850-9_1,

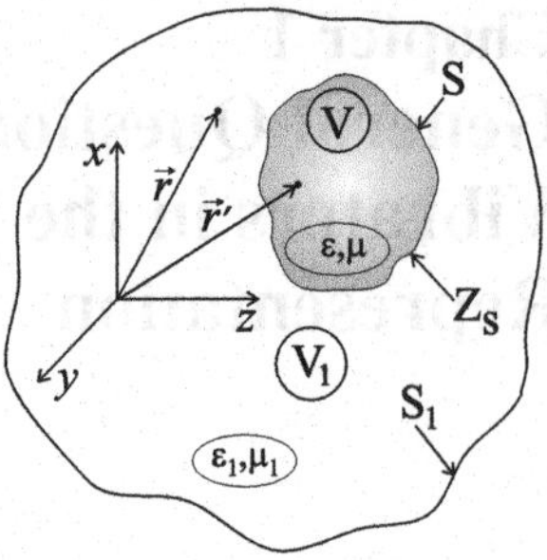

Fig. 1.1 The problem geometry

systems, for example, if S_1 is the cylindrical waveguide surface, while the body surface S has another symmetry type.

That is why we use the integral equations of macroscopic electrodynamics equivalent to the boundary problem, i.e., to Maxwell's equations, and boundary conditions on the body's surface S and in the electrodynamic volume to build a mathematical model for the electromagnetic processes in an antenna. The system of integral equations in the Gaussian unit system CGS may be written as [1, 2]

$$\begin{aligned} \vec{E}(\vec{r}) &= \vec{E}_0(\vec{r}) + (\text{grad div} + k^2\varepsilon_1\mu_1)\vec{\Pi}^{\text{e}}(\vec{r}) - ik\mu_1 \,\text{rot}\,\vec{\Pi}^{\text{m}}(\vec{r}), \\ \vec{H}(\vec{r}) &= \vec{H}_0(\vec{r}) + (\text{grad div} + k^2\varepsilon_1\mu_1)\vec{\Pi}^{\text{e}}(\vec{r}) + ik\varepsilon_1 \,\text{rot}\,\vec{\Pi}^{\text{m}}(\vec{r}). \end{aligned} \tag{1.1}$$

Here $k = 2\pi/\lambda$ is the wave number, λ is the wavelength in free space, $\vec{\Pi}^{\text{e}}(\vec{r})$ and $\vec{\Pi}^{\text{m}}(\vec{r})$ are the electrical and magnetic components of the vector Hertz potentials, equal to

$$\begin{aligned} \vec{\Pi}^{\text{e}}(\vec{r}) &= \frac{(\varepsilon/\varepsilon_1 - 1)}{4\pi} \int_V \hat{G}^{\text{e}}(\vec{r},\vec{r}\,')\vec{E}(\vec{r}\,')\,\text{d}\vec{r}\,', \\ \vec{\Pi}^{\text{m}}(\vec{r}) &= \frac{(\mu/\mu_1 - 1)}{4\pi} \int_V \hat{G}^{\text{m}}(\vec{r},\vec{r}\,')\vec{H}(\vec{r}\,')\,\text{d}\vec{r}\,', \end{aligned} \tag{1.2}$$

$\hat{G}^{\text{e}}(\vec{r},\vec{r}\,')$ and $\hat{G}^{\text{m}}(\vec{r},\vec{r}\,')$ are the electrical and magnetic tensor Green's functions for the vector potential satisfying the Helmholtz vector equation and boundary conditions on the surface S_1. Let us note that if the surface S_1 is moved away to the infinity, the boundary condition for $\hat{G}^{\text{e,m}}(\vec{r},\vec{r}\,')$ is transformed into the Sommerfeld radiation condition.

The fields on the left-hand side in (1.1) may be interpreted depending on the position of the observation point $\vec{r}$. If $\vec{r}$ belongs to the volume V, the fields $\vec{E}(\vec{r})$ and $\vec{H}(\vec{r})$ are internal fields in the body, and (1.1) are the inhomogeneous linear Fredholm integral equations of the second kind, known to have a unique solution. It should be noted once more that Maxwell's equations are partial differential equations with an infinite number of solutions, but only one solution satisfies the boundary conditions on the body's surface, and it coincides with the solution of the integral equations (1.1) [1]. If $\vec{r}$ lies outside the volume V, (1.1) become equalities, defining the full electromagnetic field in the medium outside the material body by fields of impressed sources. Thus, these equalities solve the problem of electromagnetic waves scattering (radiation) by bodies with finite dimensions if the fields inside those bodies are known. Of course, these fields may be found only if the integral equations can be solved.

Thus, solving integral equations (1.1) for all space is naturally divided into two stages. In the first stage, internal fields generated by impressed sources in the volume V are determined. In the second stage, scattered (radiated) fields are calculated by known internal fields at any point outside the volume V.

For the solution of some problems, it is convenient to express electromagnetic fields in the volume V_1 in terms of the tangential components of the field on the surface S bounding the volume V. In this case, (1.1) may be converted to the Kirchhoff–Kotler surface integral equations [1, 2]:

$$\begin{aligned}\vec{E}(\vec{r}) &= \vec{E}_0(\vec{r}) + \frac{1}{4\pi i k \varepsilon_1}(\text{grad}\,\text{div} + k_1^2)\int_S \hat{G}^{\text{e}}(\vec{r},\vec{r}')[\vec{n},\vec{H}(\vec{r}')]\,\text{d}\vec{r}' \\ &\quad - \frac{1}{4\pi}\text{rot}\int_S \hat{G}^{\text{m}}(\vec{r},\vec{r}')[\vec{n},\vec{E}(\vec{r}')]\,\text{d}\vec{r}', \\ \vec{H}(\vec{r}) &= \vec{H}_0(\vec{r}) + \frac{1}{4\pi i k \mu_1}(\text{grad}\,\text{div} + k_1^2)\int_S \hat{G}^{\text{m}}(\vec{r},\vec{r}')[\vec{n},\vec{E}(\vec{r}')]\,\text{d}\vec{r}' \\ &\quad + \frac{1}{4\pi}\text{rot}\int_S \hat{G}^{\text{e}}(\vec{r},\vec{r}')[\vec{n},\vec{H}(\vec{r}')]\,\text{d}\vec{r}'.\end{aligned} \tag{1.3}$$

Here $k_1 = k\sqrt{\varepsilon_1\mu_1}$, and $\vec{n}$ is the outward normal to the surface S.

The presentation (1.3) is used for the solution of problems in electrodynamics when the field on the surface of the material body is defined by some additional physical considerations. Thus, for good conducting bodies ($\sigma \to \infty$), the induced current is concentrated near the body's surface. Then, neglecting the thickness of the skin layer, it is possible to use the Leontovich–Schukin approximate impedance boundary condition [3]:

$$[\vec{n},\vec{E}(\vec{r})] = \overline{Z}_S(\vec{r})[\vec{n},[\vec{n},\vec{H}(\vec{r})]], \tag{1.4}$$

where $\overline{Z}_S(\vec{r}) = \overline{R}_S(\vec{r}) + i\overline{X}_S(\vec{r}) = Z_S(\vec{r})/Z_0$ is the distributed surface impedance (normalized by the characteristic impedance of free space $Z_0 = 120\pi$ ohm). Note that the impedance may vary along the body's surface.

If the observation point $\vec{r}$ lies on the body's surface S, then (1.3) and (1.4) allow us to write the integral equation for the current as

$$Z_S(\vec{r})\vec{J}^e(\vec{r}) = \vec{E}_0(\vec{r}) + \frac{1}{i\omega\varepsilon_1}(\text{grad div} + k_1^2)\int_S \hat{G}^e(\vec{r},\vec{r}')\vec{J}^e(\vec{r}')\,d\vec{r}'$$
$$+\frac{1}{4\pi}\text{rot}\int_S \hat{G}^m(\vec{r},\vec{r}')Z_S(\vec{r}')[\vec{n},\vec{J}^e(\vec{r}')]\,d\vec{r}', \qquad (1.5)$$

where $\vec{J}^e(\vec{r})$is the density of the surface electrical current

$$\vec{J}^e(\vec{r}) = \frac{c}{4\pi}[\vec{n},\vec{H}(\vec{r})] \qquad (1.6)$$

and $c \approx 2.998 \times 10^{10}$ cm/s is the velocity of light in vacuum.

Thus, the problem of scattering of electromagnetic waves (radiation) by an impedance body of finite dimensions is formulated as a rigorous boundary value problem, and is reduced to the integral equation for the surface current. Its solution is an independent problem, significant in its own right, since it offers considerable mathematical difficulties. If the characteristic dimensions of an object are much greater than the wavelength (high-frequency region), a solution is usually searched for as a series expansion in ascending powers of the inverse wave number. If the dimensions are less than the wavelength (low-frequency or quasistatic region), the representation of the unknown functions as a series expansion in powers of the wave number reduces the problem to a sequence of electrostatic problems. In contrast to the asymptotic cases, the resonant region (at least one object dimensions is comparable with the wavelength) is the most complex for analysis, requiring a rigorous solution of the field equations. It should be noted that from a practical point of view, the resonant region is of exceptional interest for thin impedance vibrators.

It should be emphasized that the choice of the absolute Gaussian unit system in this book, generally used in theoretical physics, is well justified. In the SI system, the four vectors $\vec{E}$, $\vec{D}$, $\vec{H}$ and $\vec{B}$ in Maxwell's equations, defining the field in each point of space, have different dimensions. Here $\vec{D}$ and $\vec{B}$ are the vectors of electrical and magnetic inductions, respectively. In vacuum, $\vec{D} = \varepsilon_0\vec{E}$ and $\vec{B} = \mu_0\vec{H}$, and the proportionality constants ε_0 and μ_0 play the role of permittivity and permeability, characterizing the electromagnetic characteristics of vacuum. These coefficients are included in all relations for the fields without any real sense load, and it is quite inconvenient. Certainly, it is always possible to convert CGS units to SI units and vice versa (see Appendix E).

1.2 Green's Function as the Kernel of Integral Equations

A key problem in scattering (radiation) of electromagnetic waves by material bodies of finite dimensions in a linear medium is obtaining the solution for excitation by a point source. Finding the Green's function is a classical problem of macroscopic electrodynamics, allowing an analytical representation for the solution of boundary problems. As is well known, for vector fields, Green's function is a tensor function (a symmetric tensor of the rank 2, i.e., an affinor) depending on the relative positions of an observation point $\vec{r}$ and a source point $\vec{r}'$. The tensor Green's functions in electrodynamics were introduced in [4] and have been investigated by many authors (see, for example, [5–10]).

The Green's functions for inhomogeneous vector equations, equivalent to the Maxwell's equations, are the solution of one of the following tensor equations:

$$\operatorname{rot}\operatorname{rot}\hat{G}_{\mathrm{E}}(\vec{r},\vec{r}') - k_1^2\hat{G}_{\mathrm{E}}(\vec{r},\vec{r}') = 4\pi\hat{\mathrm{I}}\delta(|\vec{r}-\vec{r}'|), \tag{1.7}$$

$$\Delta\hat{G}_{\mathrm{A}}(\vec{r},\vec{r}') + k_1^2\hat{G}_{\mathrm{A}}(\vec{r},\vec{r}') = -4\pi\hat{\mathrm{I}}\delta(|\vec{r}-\vec{r}'|), \tag{1.8}$$

which satisfy boundary conditions for the given boundary value problem. Here $\delta(|\vec{r}-\vec{r}'|) = \delta(x-x')\delta(y-y')\delta(z-z')$ is the three-dimensional Dirac delta function in Cartesian coordinates; $\hat{\mathrm{I}} = (\vec{e}_x \otimes \vec{e}_{x'}) + (\vec{e}_y \otimes \vec{e}_{y'}) + (\vec{e}_z \otimes \vec{e}_{z'})$ is the unit affinor with the unit vectors $\vec{e}_x$, $\vec{e}_y$, $\vec{e}_z$; Δ is the Laplace operator; and the symbol $\otimes$ denotes tensor product.

The Green's function allows one to obtain closed-form expressions for the electromagnetic fields such as (1.1) for an arbitrary vector source at any point in space. If a volume distribution of electrical current density $\vec{J}^{\mathrm{e}}(\vec{r})$ is specified as a source, the electric field may be represented as

$$\vec{E}(\vec{r}) = \frac{k_1^2}{\mathrm{i}\omega}\int_V \hat{G}_{\mathrm{E}}(\vec{r},\vec{r}')\vec{J}^{\mathrm{e}}(\vec{r}')\,\mathrm{d}\vec{r}', \tag{1.9}$$

or

$$\vec{E}(\vec{r}) = \frac{1}{\mathrm{i}\omega}(\operatorname{grad}\operatorname{div} + k_1^2)\int_V \hat{G}_{\mathrm{A}}(\vec{r},\vec{r}')\vec{J}^{\mathrm{e}}(\vec{r}')\,\mathrm{d}\vec{r}', \tag{1.10}$$

where $\hat{G}^{\mathrm{e}}(\vec{r},\vec{r}') = \hat{G}_{\mathrm{A}}(\vec{r},\vec{r}')$ and $\vec{E}_0(\vec{r}) = 0$ is chosen for simplicity. The two representations may be formally explained by different potential gauges. Since Maxwell's equations may be reduced to the integral equations for a field or potential, two different Green's functions may be introduced, namely $\hat{G}_{\mathrm{E}}$ for the field, and $\hat{G}_{\mathrm{A}}$ for the vector potential. It is not difficult to show that

$$\hat{G}_{\mathrm{E}} = \left(\hat{\mathrm{I}} + \frac{1}{k_1^2}\{\operatorname{grad}\otimes\operatorname{grad}\}\right)\hat{G}_{\mathrm{A}} \tag{1.11}$$

if the operation $\{\operatorname{grad}\otimes\operatorname{grad}\}$ is interpreted as the tensor product of two symbolic vectors.

The only boundary condition imposed on the Green's function for infinite space is the Sommerfeld radiation condition, and the Green's function may be represented as

$$\hat{G}_{\mathrm{A}}(\vec{r},\vec{r}\,') = \hat{\mathrm{I}}\frac{\mathrm{e}^{-ik_1|\vec{r}-\vec{r}\,'|}}{|\vec{r}-\vec{r}\,'|}. \tag{1.12}$$

The Green's function (1.12) defines the diverging waves if time-dependence is chosen as $\mathrm{e}^{i\omega t}$. For waves approaching the source (the converging waves), substitution $-k_1$ for k_1 in (1.12) is necessary. When a wave is reflected back to the source, a combination of the diverging and the converging waves must be used.

If $\vec{r}=\vec{r}\,'$, the functions $\hat{G}_{\mathrm{A}}$and $\hat{G}_{\mathrm{E}}$ in (1.12) and (1.11) tend to infinity, and it is impossible to define the integrals in a general sense as the limit of the integral sum, since it does not exist. So, strictly speaking, (1.9) and (1.10) are valid for those points of space where the sources are absent. When the observation point $\vec{r}$ coincides with any point $\vec{r}\,'$ where the source is present, the volume integrals in (1.9) and (1.10) must be treated as improper, that is, $\int\limits_V = \lim\limits_{\rho\to 0}\int\limits_{V-v}$ where v is the eliminated volume contained in a sphere with infinitesimally small radius ρ centered at $\vec{r}\,'$. The difference between the integrals is that the improper integral (1.10) can be converted into an absolutely convergent integral [8], while the integral in (1.9) is principally conditionally convergent [8], and its value depends on the shape of the eliminated region containing the singular point $\vec{r}=\vec{r}\,'$. To obtain a physically correct result, it is necessary to integrate (1.9) in terms of the integral principal value, defining the main contribution to the integration result. The integrals may also be evaluated as non-singular by expanding the singular part in Green's functions. The behavior of the tensor Green's function (1.11) in a neighborhood of source points is very difficult to investigate. Any more rigorous treatment requires using generalized functions [9]. In this connection, it should be noted that the expression for the field (1.10) is preferable to (1.9) and (1.11), since (1.10) has an integrated singularity, and calculation of the principal integral value is avoided, though parametric differentiation under the integral sign in (1.10) becomes necessary.

This reasoning is completely applicable for bounded regions, since the Green's function for the bounded region has the same singularity order as for unbounded space [6]. It should be noted once more that the Green's function for a bounded region is defined as the solution of (1.7) or (1.8), defining the field (1.9) or (1.10), satisfying boundary conditions on the surface S_1. Thus, the general solution of (1.8) can be represented as [1]:

$$\hat{G}(\vec{r},\vec{r}\,') = \hat{\mathrm{I}}G(\vec{r},\vec{r}\,') + \hat{G}_0^{V_1}(\vec{r},\vec{r}\,'), \tag{1.13}$$

where $G(\vec{r},\vec{r}\,') = \mathrm{e}^{-ik_1|\vec{r}-\vec{r}\,'|}/|\vec{r}-\vec{r}\,'|$, and $\hat{G}_0^{V_1}(\vec{r},\vec{r}\,')$ is a regular function satisfying the homogeneous equation

$$\Delta\hat{G}_0^{V_1}(\vec{r},\vec{r}\,') + k_1^2\hat{G}_0^{V_1}(\vec{r},\vec{r}\,') = 0 \tag{1.14}$$

and satisfying, in conjunction with $\hat{I}G(\vec{r},\vec{r}')$, boundary conditions on the surface S_1, enclosing the volume V_1, for the field of the point source located at the point $\vec{r}'$. If S_1 is a perfectly conducting infinite plane, the Green's function (1.13) can be constructed by the mirror-image or eigenfunction method [6], equivalent from a physical viewpoint, but somewhat different in mathematical realization. The advantage of the mirror-image method is its clarity, especially when V_1 is a half-space over a perfectly conducting screen. However, this method for some other configuration of the volume V_1 may be not so simple, and practical calculations lead in many cases to rather complex analytical expressions. The eigenfunction method is more general, and is applicable for various shapes of the volume V_1, for example for cylindrical waveguides and resonators, spherical resonators, unbounded space outside the perfectly conducting sphere, and so on (see Appendix A).

1.3 Integral Equations for a Current on Thin Impedance Vibrators

A direct solution of (1.5) for the material body V with arbitrary surface may result in mathematical difficulties. However, the problem is considerably simplified for impedance cylinders if the cross-sectional perimeter is less than their lengths and the wavelength in the medium (thin vibrators). Moreover, in this case it is possible to extend the boundary condition (1.4) on the cylindrical surfaces to an arbitrary complex impedance distribution regardless of the exciting field structure and the electrophysical characteristics of the vibrator material.

Let us transform the integral equation (1.5) into a form applicable to a thin vibrator consisting of a bounded circular cylindrical wire with radius r and length $2L$ (in the general case it may have a curvilinear axial configuration), satisfying the inequalities

$$\frac{r}{2L} \ll 1, \quad \left|\frac{r}{\lambda_1}\right| \ll 1, \quad \frac{r}{\tilde{r}} \ll 1, \tag{1.15}$$

where λ_1 is the wavelength in the medium and $\tilde{r}$ is curvature radius of the vibrator axial line. These inequalities make it possible to express the density of an induced current having only a longitudinal component (we omit the index "e") as

$$\vec{J}(\vec{r}) = \vec{e}_s J(s)\psi(\rho,\varphi), \tag{1.16}$$

distributed over the cross section of the vibrator as in a quasistationary case [11]. For the function $\psi(\rho,\varphi)$, the relation

$$\int_{\perp} \psi(\rho,\varphi)\rho\, \mathrm{d}\rho\, \mathrm{d}\varphi = 1 \tag{1.17}$$

holds. In (1.16) and (1.17), $\vec{e}_s$ is the unit vector along the tangent to the axis $\{0s\}$ associated with the vibrator; $\psi(\rho,\varphi)$ is the function of the transverse ($\perp$) polar coordinates ρ and φ; $J(s)$ is the unknown current satisfying the boundary conditions on the ends of the vibrator:

$$J(-L) = J(L) = 0. \tag{1.18}$$

Projecting (1.5) onto the vibrator axis and taking into account that $[\vec{n}, \vec{J}^e(\vec{r})] \ll 1$ due to (1.15), we obtain the equation for the current in a thin impedance vibrator located in a homogeneous isotropic infinite medium:

$$z_i(s)J(s) = E_{0s}(s) + \frac{1}{i\omega\varepsilon_1}\int_{-L}^{L}\left[\frac{\partial}{\partial s}\frac{\partial J(s')}{\partial s'} + k_1^2(\vec{e}_s\vec{e}_{s'})J(s')\right]G_s(s,s')\,ds'. \tag{1.19}$$

Here $E_{0s}(s)$ is the projection of the impressed field parallel to the vector $\vec{e}_s$, $z_i(s)$ is the internal impedance per unit length of the vibrator in ohm/m, $(Z_S(\vec{r}) = 2\pi r z_i(\vec{r}))$, $\vec{e}_{s'}$ is the unit vector of the axis $\{0s\}$ associated with the vibrator surface, and

$$G_s(s,s') = \int_{-\pi}^{\pi}\frac{e^{-ik_1\sqrt{(s-s')^2+[2r\sin(\varphi/2)]^2}}}{\sqrt{(s-s')^2+[2r\sin(\varphi/2)]^2}}\psi(r,\varphi)r\,d\varphi \tag{1.20}$$

is the exact kernel of the integral equation.

Serious difficulties may be encountered in solving (1.19) with the kernel (1.20). Therefore, the thin-wire approximation is usually used [11]:

$$G_s(s,s') = \frac{e^{-ik_1R(s,s')}}{R(s,s')}, \quad R(s,s') = \sqrt{(s-s')^2 + r^2}. \tag{1.21}$$

Here the source points are located on the geometric axis of the vibrator, and the observation points on its physical surface. The function $G_s(s,s')$ is continuous everywhere, and the equation for the current is sufficiently simplified without noticeable loss of accuracy [12]. For the rectilinear conductor $((\vec{e}_s\vec{e}_{s'}) = 1)$, integration by parts in (1.19) with the boundary conditions (1.18) results in the equation

$$\left(\frac{d^2}{ds^2} + k_1^2\right)\int_{-L}^{L} J(s')G_s(s,s')\,ds' = -i\omega\varepsilon_1 E_{0s}(s) + i\omega\varepsilon_1 z_i(s)J(s). \tag{1.22}$$

If the vibrator is located in the bounded electrodynamic volume (a waveguide or a resonator), we obtain, with (1.5) and (1.13),

$$\int_{-L}^{L}\left\{\left[\frac{\partial}{\partial s}\frac{\partial J(s')}{\partial s'}+k_1^2(\vec{e}_s\vec{e}_{s'})J(s')\right]G_s(s,s')+J(s')\left[\frac{\partial^2}{\partial s^2}+k_1^2\right]\vec{e}_s(\hat{G}_{0s}^{V_1}(s,s')\vec{e}_{s'})\right\}ds'$$
$$=-i\omega\varepsilon_1E_{0s}(s)+i\omega\varepsilon_1z_i(s)J(s), \tag{1.23}$$

where $G_{0s}^{V_1}(s,s')=\int\limits_{\perp}G_{0s}^{V_1}(s,\rho,\phi;s',\rho',\phi')\psi(\rho',\phi')\rho'\,d\rho'\,d\phi'$. Then for a rectilinear impedance vibrator located in the volume V_1, the equation for the current has the form

$$\left(\frac{d^2}{ds^2}+k_1^2\right)\int_{-L}^{L}J(s')G_s(s,s')\,ds'=-i\omega\varepsilon_1E_{0s}(s)+i\omega\varepsilon_1z_i(s)J(s)-F_0[s,J(s)], \tag{1.24}$$

where

$$F_0[s,J(s)]=\left(\frac{d^2}{ds^2}+k_1^2\right)\int_{-L}^{L}J(s')G_{0s}^{V_1}(s,s')\,ds'. \tag{1.25}$$

Equation (1.22) is known as Pocklington's integral equation [13, 14] when the volume V_1 is free space and $z_i=0$, while (1.19), after integration over S, is known as Mei's equation [15].

The integral equations (1.3), (1.5), (1.19), (1.22)–(1.24) are formulated for a single body, but they can be easily generalized to a system of bodies by the superposition principle. Here the domain of integration is the surfaces of all the bodies, and the currents on any surface can be found by solving a system of integral equations.

Let us note in conclusion that one of the most universal numerical methods for solving the above-mentioned integral equations is the method of moments (see Appendix B). A detailed description applicable to thin perfectly conducting vibrators can be found in many references; see, for example, [16]. However, in this book we shall use approximate analytical methods as more illustrative than the analogous numeric methods. That is why we shall use the method of moments only for comparative analysis of data obtained by both the numerical and the approximate analytical methods.

1.4 Approximate Analytical Methods for the Solution of Integral Equations

The rigorous solution of the above integral equations for the electrical current in an impedance vibrator cannot be evaluated in closed form. But it should not be supposed that the current distribution cannot be approached by an approximate solution with a good degree of accuracy [17]. It is a good idea to use methods developed earlier for perfectly conducting vibrators. Thus for rectilinear vibrators in free space, there exist several time-tested techniques, namely the successive iterations method [11, 18], series expansion of the unknown function in a small parameter [19], the variation procedure [11], and the "key equation" method [20]. To determine the advantages and disadvantages of various approximate analytical methods, we solve (1.24) for the electromagnetic wave incident on a perfectly conducting vibrator ($z_i = 0$), located in the volume V_1 with $\varepsilon_1 = \mu_1 = 1$, by a series expansion in a small parameter for an unknown function (referred to as the small-parameter method) and the successive iterations method (referred to as the iterations method).

1.4.1 Series Expansion Technique

The initial equation for analysis may be presented as

$$\left(\frac{d^2}{ds^2}+k^2\right)\int\limits_{-L}^{L} J(s')\frac{e^{-ikR(s,s')}}{R(s,s')}ds' = -i\omega E_{0s}(s) - f_0[s,J(s)], \tag{1.26}$$

where $R(s,s') = \sqrt{(s-s')^2 + r^2}$, and

$$f_0[s,J(s)] = \left(\frac{d^2}{ds^2}+k^2\right)\int\limits_{-L}^{L} J(s')G_{0s}^{V_1}(s,s')\,ds' \tag{1.27}$$

is the regular part of the vibrator field depending on the geometry of the volume V_1.

A change of variables in (1.26), in view of

$$\left.\begin{aligned} s' &= s - \sqrt{R^2 - r^2}, \quad \text{if } s' \leqslant s \\ s' &= s + \sqrt{R^2 - r^2}, \quad \text{if } s' \geqslant s \end{aligned}\right\},$$

transforms (1.26) to

$$\begin{aligned} &\left(\frac{d^2}{ds^2}+k^2\right)\left\{-\int\limits_{-L}^{s} J(s')e^{-ikR}d\ln[C(R+\sqrt{R^2-r^2})]\right. \\ &\left.+\int\limits_{s}^{l} J(s')e^{-ikR}d\ln[C(R+\sqrt{R^2-r^2})]\right\} = -i\omega E_{0s}(s) - f_0[s,J(s)], \end{aligned} \tag{1.28}$$

where C is an arbitrary constant. By partial integration in (1.28), and taking into account the boundary conditions for the current (1.18) we obtain

$$\left(\frac{\mathrm{d}^2}{\mathrm{d}s^2}+k^2\right)\left\{J(s)\mathrm{e}^{-\mathrm{i}kr}\ln Cr+\int\limits_{-L}^{s}\ln[C(R+\sqrt{R^2-r^2})]\,\mathrm{d}[J(s')\mathrm{e}^{-\mathrm{i}kR}]\right.$$
$$\left.-\int\limits_{s}^{L}\ln[C(R+\sqrt{R^2-r^2})]\,\mathrm{d}[J(s')\mathrm{e}^{-\mathrm{i}kR}]\right\}=\mathrm{i}\omega E_{0s}(s)+f_0[s,J(s)]. \tag{1.29}$$

Assuming, in view of (1.15), $e^{-ikr}=1$ and choosing $C=1/2L$ (note that in [19], $C=k$), we reduce (1.26) to the following integrodifferential equation for the current:

$$\frac{\mathrm{d}^2J(s)}{\mathrm{d}s^2}+k^2J(s)=\alpha\{\mathrm{i}\omega E_{0s}(s)+f[s,J(s)]+f_0[s,J(s)]\}. \tag{1.30}$$

Here $\alpha=1/(2\ln[r/(2L)])$ is a small parameter, and

$$f[s,J(s)]=-\left(\frac{\mathrm{d}^2}{\mathrm{d}s^2}+k^2\right)\int\limits_{-L}^{L}\mathrm{sign}(s-s')\ln\frac{R+(s-s')}{2L}\frac{\mathrm{d}}{\mathrm{d}s'}[J(s')\mathrm{e}^{-\mathrm{i}kR}]\,\mathrm{d}s' \tag{1.31}$$

is the self-field of the vibrator in free space.

Let us represent $J(s)$ expanding it into power series in the small parameter $|\alpha|\ll 1$:

$$J(s)=J_0(s)+\alpha J_1(s)+\alpha^2J_2(s)+\cdots. \tag{1.32}$$

Substituting (1.32) into (1.27) and (1.31), we derive the series

$$f_\Sigma[s,J(s)]=f_\Sigma[s,J_0(s)]+\alpha f_\Sigma[s,J_1(s)]+\alpha^2 f_\Sigma[s,J_2(s)]+\cdots, \tag{1.33}$$

where $f_\Sigma[s,J(s)]=f[s,J(s)]+f_0[s,J(s)]$ is the total self-field of the vibrator. Substituting (1.32) and (1.33) into (1.30) and equating multipliers with equal powers of α on the left-hand and the right-hand sides of the equation now yields a system of ordinary differential equations:

$$\begin{aligned}
&\frac{\mathrm{d}^2J_0(s)}{\mathrm{d}s^2}+k^2J_0(s)=0,\\
&\frac{\mathrm{d}^2J_1(s)}{\mathrm{d}s^2}+k^2J_1(s)=\mathrm{i}\omega E_{0s}(s)+f_\Sigma[s,J_0(s)],\\
&\frac{\mathrm{d}^2J_2(s)}{\mathrm{d}s^2}+k^2J_2(s)=f_\Sigma[s,J_1(s)],\\
&\cdots\cdots\cdots\cdots\cdots\cdots\cdots\cdots\cdots\cdots\\
&\frac{\mathrm{d}^2J_n(s)}{\mathrm{d}s^2}+k^2J_n(s)=f_\Sigma[s,J_{n-1}(s)],
\end{aligned} \tag{1.34}$$

which can be solved by the method of successive approximations. Each equation of the system is solved subjected to the boundary conditions as in (1.18), namely $J_0(\pm L) = 0, \quad J_1(\pm L) = 0, \quad J_2(\pm L) = 0, \ldots, J_n(\pm L) = 0$.

The solution of first equation in the system (1.34) does not depend on the exciting field $E_{0s}(s)$,

$$J_0(s) = C_1 \cos ks + C_2 \sin ks, \tag{1.35}$$

which satisfies the boundary conditions only when the relations

$$C_1 = 0 \text{ at } 2L = m\lambda \text{ and } C_2 = 0 \text{ at } 2L = (2n+1)\frac{\lambda}{2} \tag{1.36}$$

are satisfied. Here m and n are integers. If the vibrator length $2L$ does not satisfy (1.36), then $J_0 \equiv 0$, $f_\Sigma[s, J_0(s)] \equiv 0$, and the current to a first approximation is equal to

$$J(s) = \alpha J_1(s) = -\alpha \frac{i\omega/k}{\sin 2kL} \left\{ \sin k(L-s) \int_{-L}^{s} E_{0s}(s') \sin k(L+s')\,ds' + \sin k(L+s) \int_{s}^{L} E_{0s}(s') \sin k(L-s')\,ds' \right\}. \tag{1.37}$$

As can be seen, the expression for the current does not include the functions depending upon vibrator own field, $f[s, J(s)]$ and $f_0[s, J(s)]$, which define the vibrator's resonance and energy characteristics. Obviously, it is necessary to obtain further approximations in order to take account of $f_\Sigma[s, J(s)]$, but due to essential mathematical difficulties, only $J_2(0)$ for the vibrator fed at the center in free space is known [19].

As an example, let us consider a scattering problem by a vibrator located in the cross-sectional plane of a standard rectangular waveguide, parallel to its narrow wall. The impressed field is equal to

$$E_{0s}(s) = E_0 \sin \frac{\pi x_0}{a}. \tag{1.38}$$

Here E_0 is the amplitude of the H_{10} mode, incident from the region $z = -\infty$, and x_0 is the distance from the waveguide's narrow wall to the vibrator's axial line. Then the current induced on the vibrator equals (1.37):

$$J(s) = -\alpha E_0 \sin \frac{\pi x_0}{a} \frac{i\omega}{k^2} \frac{(\cos ks - \cos kL)}{\cos kL}. \tag{1.39}$$

And finally, we present the solution derived by the small parameter method to first approximation for the classical problem of the normal incidence of a plane electromagnetic wave on the vibrator in free space ($E_{0s}(s) = E_0$):

$$J(s) = -\alpha E_0 \frac{i\omega}{k^2} \frac{(\cos ks - \cos kL)}{\cos kL}. \tag{1.40}$$

We note in conclusion that for (1.39) and (1.40), the usability condition [19] becomes

$$\left| kL - n\frac{\pi}{2} \right| \gg |\alpha|, \tag{1.41}$$

which in conjunction with (1.36) limits essentially the applicability of this solution.

1.4.2 Successive Iterations Method

The successive iterations method was suggested by Hallen [18] and improved by King [11]. Let us use this method to investigate the characteristics of vibrators in free space in order to eliminate the above-mentioned disadvantages of the solution of the integral equation by the small parameter method.

Inverting the differential operator on the left-hand side of (1.26), we obtain the following integral equation:

$$\int_{-L}^{L} J(s')G_s^{\Sigma}(s,s')\,ds' = C_1 \cos ks + C_2 \sin ks - \frac{i\omega}{k}\int_{-L}^{s} E_{0s}(s')\sin k(s-s')\,ds', \tag{1.42}$$

where $G_s^{\Sigma}(s,s') = G_s(s,s') + G_{0s}^{V_1}(s,s')$. To find one of the arbitrary constants C_1 and C_2 connected with the method of vibrator excitation, symmetry conditions [11] must be applied. In other words, at this stage of the solution by the successive iterations method, the field of the impressed sources $E_{0s}(s)$ must be specified. Assume, in accordance with (1.38), $E_{0s}(s) = E_0$, which corresponds to vibrator excitation by the main mode in a rectangular waveguide, when the vibrator axis is located at $x_0 = a/2$. Then

$$\int_{-L}^{L} J(s')G_s^{\Sigma}(s,s')\,ds' = C_1 \cos ks + \frac{i\omega}{k^2}E_0(\cos ks \cos kL - 1). \tag{1.43}$$

Note that (1.43) is analogous to Hallen's linearized integral equation [11, 18], which is the basis for many publications dealing with the theory of thin vibrator antennas for $G_s^{\Sigma}(s,s') = e^{-ikR}/R$.

The kernel of the integral equation (1.43) has on the vibrator's surface a singularity of quasistationary type. Let us isolate it using (1.15) and rewriting the left-side of (1.43) as

$$\int_{-L}^{L} J(s')G_s^{\Sigma}(s,s')\,ds' = \int_{-L}^{L} J(s')\frac{e^{-ikR(s,s')}}{R(s,s')}\,ds' + \int_{-L}^{L} J(s')G_{0s}^{V_1}(s,s')\,ds'. \quad (1.44)$$

Then

$$\int_{-L}^{L} J(s')\frac{e^{-ikR(s,s')}}{R(s,s')}\,ds' = \Omega(s)J(s) + \int_{-L}^{L}\left[J(s')\frac{e^{-ikR(s,s')}}{R(s,s')} - \frac{J(s)}{R(s,s')}\right]ds', \quad (1.45)$$

where

$$\Omega(s) = \int_{-L}^{L}\frac{ds'}{\sqrt{(s-s')^2+r^2}}. \quad (1.46)$$

The first summand on the right-hand side of (1.45) is logarithmically large in comparison to the second regular term, and the function $\Omega(s)$ differs from its average value $\overline{\Omega}(s) = 2\ln(2L/r) - 0.614$ only at the vibrator's ends, where the current vanishes, since $J(\pm L) = 0$. Thus we may transform (1.43) into

$$J(s) = -\alpha\left[C_1\cos ks + \frac{i\omega}{k^2}E_0(\cos ks\cos kL - 1)\right] + \alpha\int_{-L}^{L}\left[J(s')G_s^{\Sigma}(s,s') - \frac{J(s)}{R(s,s')}\right]ds', \quad (1.47)$$

where $\alpha = 1/(2\ln[r/(2L)])$ is the small parameter coinciding with that obtained above in Sect. 1.4.1 with the integration constant $C = 1/2L$.

Following the procedure described in [11, 18], assume $s = L$ in (1.47) and subtract the obtained result from (1.47) (in fact, the subtrahend is equal to zero, since $J(L) \equiv 0$). Equation (1.47) is now transformed to

$$J(s) = -\alpha\left[C_1(\cos ks - \cos kL) + \frac{i\omega}{k^2}E_0\cos kL(\cos ks - \cos kL)\right]$$
$$+\alpha\left\{\int_{-L}^{L}\left[J(s')G_s^{\Sigma}(s,s') - \frac{J(s)}{R(s,s')}\right]ds' - \int_{-L}^{L} J(s')G_s^{\Sigma}(L,s')\,ds'\right\}. \quad (1.48)$$

Choosing the first term of (1.48) as the zero-order approximation to the current $J_0(s)$ and using (1.18) to define the constant C_1, we obtain

$$J_0(s) = -\alpha E_0\frac{i\omega}{k^2}\frac{(\cos ks - \cos kL)}{\cos kL}, \quad (1.49)$$

coinciding with (1.39) for $x_0 = a/2$ obtained by the small parameter method as a first approximation. Substituting (1.49) into (1.48), we obtain a first approximation for the current with accuracy of order α^2:

$$J_1(s) = -\alpha E_0 \frac{\mathrm{i}\omega}{k^2} \frac{(\cos ks - \cos kL)}{\cos kL + \alpha F(kr, kL)}, \tag{1.50}$$

where

$$F(kr, kL) = \int_{-L}^{L} \left[(\cos ks' - \cos kL) G_s^{\Sigma}(L, s')\right] \mathrm{d}s' \tag{1.51}$$

is the function of the vibrator's self-field, allowing analysis by a single formula for both tuned ($\cos kL = 0$) and untuned ($\cos kL \neq 0$) vibrators even to a first approximation in α (in contrast to the small parameter method). If the vibrator is in free space $G_s^{\Sigma}(s, s') = \mathrm{e}^{-ikR}/R$, the integral in (1.51) can be evaluated analytically by generalized integral functions [11] (see Appendix C).

Thus the solution of a quasi-unidimensional integral equation for the electrical current in thin vibrators by the small parameter method leads to different expressions for the current in tuned (the frequency of the impressed field differs little from the intrinsic frequency of the vibrator) and in untuned vibrators (when this condition is not true), though the solution for the untuned vibrator and arbitrary excitation can be derived to a first approximation. The solution of the integral equation for the current by the iterations method is given by a single formula that is suitable for both tuned and untuned vibrators. However, this method is applicable only when the fields of the impressed sources are specified at the initial stage of analysis. A general analytical expression for the current, given as a single formula and suitable both for tuned and untuned vibrators without specification of the impressed source fields and electrodynamic volumes, will be obtained by the asymptotic averaging method in the next chapter.

1.5 Averaging Method

Rigorous verification of the asymptotic averaging method is a pure mathematical problem, investigated in detail in the monographs [21, 22], where analogous theorems are proved. Let us briefly consider the main features of the method.

Let a system of ordinary differential equations in standard form be given by

$$\frac{\mathrm{d}x}{\mathrm{d}s} = \alpha X(s, x), \tag{1.52}$$

where x is an n-dimensional vector and $0<\alpha \ll 1$ is a small parameter. The proportionality of the first derivatives $\mathrm{d}x/\mathrm{d}s$ to a small parameter implies that the variables x are changing slowly. There exist many different methods to reduce the initial equations to the form (1.52), but most often, the method of arbitrary constants variation is used [22]. Considering that reduction of the initial system to standard form has already been done, we change variables in (1.52) in accordance with

$$x = \xi + \alpha\tilde{X}(s, \xi), \tag{1.53}$$

where ξ are new unknowns, $\partial\tilde{X}/\partial s = X(s, \xi) - \overline{X}(\xi)$, and the bar symbol designates averaging over the variable S:

$$\overline{X}(\xi) = \lim_{l\to\infty} \frac{1}{l} \int_0^l X(s, \xi)\,\mathrm{d}s. \tag{1.54}$$

After some transformations we obtain [21]

$$\frac{\mathrm{d}\xi}{\mathrm{d}s} = \alpha\overline{X}(\xi) + \alpha^2 \ldots, \tag{1.55}$$

that is, if ξ satisfies (1.55) with the right-hand side differing from the right-hand side of the averaged equations

$$\frac{\mathrm{d}\xi}{\mathrm{d}s} = \alpha\overline{X}(\xi) \tag{1.56}$$

by order of magnitude α^2 and (1.53) represents the exact solution of the initial equations (1.52). Therefore, we may accept $x = \xi$ as a first approximation, with ξ representing the solution of (1.56), satisfying (1.52) infinitesimally to second order. Expression (1.53) with ξ satisfying (1.55) is called the improved first approximation.

Thus, equations for the first approximation (1.56) are obtained from the exact equations (1.52) by averaging over the variable s, while the ξ are treated as constants. This formal procedure whereby the exact equations are replaced by the averaged ones is called the averaging principle. Its essence was brought to light by Bogoliubov and Mitropolsky in [21], who showed that there exists a certain change of variables that make it possible to omit s from the right-hand side of the equations to any degree of precision relative to the small parameter α. It provides the opportunity to create not only the first-approximation system (1.56) but to obtain higher-order averaged systems giving approximate solutions of (1.52) with arbitrary accuracy. However, practically, only the first approximation can work effectively due to rapid complications in the formulas. The error evaluation for the first approximation is also given in [21], where it was found that the difference $x(s) - \xi(s)$ can be made indefinitely small for sufficiently small α on an arbitrarily large but finite interval $0 < s < l$. Thus, the following main averaging theorem holds [21, 22].

Theorem. *Let functions* $X(s,x)$ *be defined and continuous in the region* $Q(s\geqslant 0, x \in D)$*. Assume that*

1. $X(s,x) \in \mathrm{Lip}_x(\lambda, Q)$*, that is,* $X(s,x)$ *is Lipschitz continuous in* x *with the constant* λ
2. $\|X(s,x)\|<M$*, i.e., the function* $X(s,x)$ *is bounded*
3. *There exists a limit (1.54) at each point* $x \in D$
4. *The averaged system solution* $\xi(s)$ *(1.56) is determined for all* $s\geqslant 0$ *and is located in the region* D *within a* ρ*-neighborhood*

Then it is possible to choose, for any indefinitely small $\eta>0$ and for arbitrarily large $L>0$, an α_0 such that the inequality $\|x(s)-\xi(s)\|<\eta$ holds on the segment $0\leqslant s\leqslant L\alpha^{-1}$ for $0<\alpha<\alpha_0$*, where* $x(s)$ *and* $\xi(s)$ *are respectively the solutions of the systems (1.52) and (1.56), coinciding for* $s=0$*.*

The theorem may be also generalized to a system of integrodifferential equations as

$$\frac{dx}{ds} = \alpha X\left(s, x, \int_0^s \phi(s, s', x(s'))\, ds'\right). \tag{1.57}$$

Here the following averaging scheme is possible [22]. Let us calculate the integral

$$\psi(s,x) = \int_0^s \phi(s, s', x)\, ds' \tag{1.58}$$

over the explicit variable s' (s and x are parameters). Let us consider along with (1.57) a system of differential equations

$$\frac{dy}{ds} = \alpha X(s, y, \psi(s,y)), \tag{1.59}$$

and let the limit

$$\lim_{l\to\infty} \frac{1}{l}\int_0^l X(s, y, \psi(s,y))\, ds = \overline{X}(y) \tag{1.60}$$

exist. Then we obtain the following system of differential equations:

$$\frac{d\xi}{ds} = \alpha \overline{X}(\xi). \tag{1.61}$$

The system (1.61) is averaged, corresponding to the system of integrodifferential equations (1.57). Thus, the averaging for the systems (1.57) consists in approximating the solutions of the initial system by a specially selected system of differential equations (1.61), which is easier to solve than the initial system of integrodifferential equations. The conditions for the solutions of (1.57), (1.59), and (1.61) to be close are given in [22].

Different averaging schemes are permitted for systems of integrodifferential equations (in contrast to systems of differential equations). Generally, a system of integrodifferential equations may be replaced by several different systems of averaged equations. Some of these averaged systems are systems of differential equations, and others are systems of integrodifferential equations. This possibility ensures the high efficiency of the averaging method for applied problems solutions.

Let us consider the system of integrodifferential equations

$$\frac{dx}{ds} = \alpha X\left(s, x, \frac{dx}{ds}, \int\limits_0^s \phi\left(s, s', x(s'), \frac{dx(s')}{ds'}\right) ds'\right) \tag{1.62}$$

in standard form unsolved relative to the derivative. An attempt to solve the system relative to dx/ds often leads to intricate and laborious calculations. But these difficulties can be obviated by special averaging schemes [22]. Thus, to a first approximation, the simplified system

$$\frac{dy}{ds} = \alpha X\left(s, y, 0, \int\limits_0^s \phi(s, s', y(s'), 0)\, ds'\right) \tag{1.63}$$

may be used instead of (1.62), since the derivatives on the right-hand side of (1.62) begin to take affect for the second and further asymptotic approximations and also for improved first approximations. Under rather general conditions and for small α, the solutions (1.62) and (1.63) are infinitely close in the segment of order $L\alpha^{-1}$.

For systems of differential and integrodifferential equations, there exist various partial averaging schemes in which only some summands or equations in the system are averaged [22]. For example, if the initial system is represented in the form

$$\frac{dx}{ds} = \alpha X_1(s, x) + \alpha X_2(s, x) \tag{1.64}$$

and the limit

$$\lim_{l \to \infty} \frac{1}{l} \int\limits_0^l X_1(s, x)\, ds = \overline{X}_1(x) \tag{1.65}$$

exists, then a partially averaged system corresponding to the system (1.64) may be represented as

$$\frac{\mathrm{d}\xi}{\mathrm{d}s} = \alpha\overline{X}_1(\xi) + \alpha X_2(s,\xi). \tag{1.66}$$

Partial averaging schemes are highly diversified, and they may be extended to the systems defined in (1.57) and (1.62).

References

1. Khizhnyak, N.A.: Integral Equations of Macroscopical Electrodynamics. Naukova dumka, Kiev (1986) (in Russian).
2. Nesterenko, M.V., Katrich, V.A., Penkin, Yu.M., Berdnik, S.L.: Analytical and Hybrid Methods in the Theory of Slot-Hole Coupling of Electrodynamic Volumes. Springer, New York (2008).
3. Leontovich, M.A.: On Approximate Boundary Conditions for the Electromagnetic Field on Surfaces of Good Conductive Bodies. Investigations of Radiowave Propagation. Printing House of the Academy of Sciences of the USSR, Moscow-Leningrad (1948) (in Russian).
4. Levin, H., Schwinger, J.: On the theory of electromagnetic wave diffraction by an aperture in an infinite plane conducting screen. Commun. Pure Appl. Math. **3**, 355–391 (1950).
5. Collin, R.E.: Field Theory of Guided Waves. McGraw-Hill, New York (1960).
6. Morse, P.M., Feshbach, H.: Methods of Theoretical Physics. McGraw-Hill, New York (1953).
7. Tai, C.T.: Dyadic Green's Function in Electromagnetic Theory. Intex Educational Publishers, Scranton (1971).
8. Tikhonov, A.N., Samarsky, A.A.: Equations of Mathematical Physics. Nauka, Moscow (1977) (in Russian).
9. Van Bladel, J.: Some remarks on Green's dyadic for infinite space. IEEE Trans. Antennas Propag. **AP-9**, 563–566 (1961).
10. Felsen, L.B., Marcuvitz, N.: Radiation and Scattering of Waves. Prentice-Hall, Inc., Englewood Cliffs, NJ, (1973).
11. King, R.W.P.: The Theory of Linear Antennas. Harvard University Press, Cambridge, MA (1956).
12. King, R.W.P., Aronson, E.A., Harrison, C.W.: Determination of the admittance and effective length of cylindrical antennas. Radio Sci. **1**, 835–850 (1966).
13. Pocklington, H.C.: Electrical oscillations in wires. Proc. Camb. Philol Soc. **9**, pt VII, 324–332 (1897).
14. Brillouin, L.: The antenna problem. Quart. Appl. Math. **1**, 201–214 (1943).
15. Mei, K.K.: On the integral equation of thin wire antennas. IEEE Trans. Antennas Propag. **AP-13**, 374–378 (1965).
16. Mittra, R. (ed.): Computer Techniques for Electromagnetics. Pergamon, New York (1973).
17. Nesterenko, M.V.: Analytical methods in the theory of thin impedance vibrators. Prog. Electromagn. Res. B **21**, 299–328 (2010).
18. Hallen, E.: Theoretical investigations into the transmitting and receiving qualities of antennas. Nova Acta Reg. Soc. Sci. Ups. Ser. IV **11**, 1–44 (1938).
19. Leontovich, M., Levin, M.: On the theory of oscillations excitation in antennas' vibrators. J. Tech. Phys. **14**, 481–506 (1944) (in Russian).
20. Vineshtein, L.A.: Current waves in a thin cylindrical conductor. J. Tech. Phys. **29**, 65–91 (1959) (in Russian).
21. Bogoliubov, N.N., Mitropolsky, Y.A.: Asymptotic Methods in the Theory of Nonlinear Oscillations. Internat. Monog. Adv. Math. Physics, Gordon and Breach, New York (1961).
22. Philatov, A.N.: Asymptotic Methods in the Theory of Differential and Integrodifferential Equations. PHAN, Tashkent (1974) (in Russian).

Chapter 2
Radiation of Electromagnetic Waves by Impedance Vibrators in Free Space and Material Medium

2.1 Asymptotic Solution of Integral Equations for Vibrator Current in Free Space

Let us rewrite (1.22) ($z_i(s) = \text{const}, \varepsilon_1 = \mu_1 = 1$), using the approximate kernel (1.21), the quasi-unidimensional analog of the exact integral equation with the kernel (1.20), as

$$\left(\frac{d^2}{ds^2} + k^2\right) \int_{-L}^{L} J(s') \frac{e^{-ikR(s,s')}}{R(s,s')} ds' = -i\omega E_{0s}(s) + i\omega z_i J(s), \tag{2.1}$$

where $R(s,s') = \sqrt{(s-s')^2 + r^2}$. It is obvious that $F_0[s, J(s)] \equiv 0$. We isolate the logarithmic kernel singularity as in (1.45):

$$\int_{-L}^{L} J(s') \frac{e^{-ikR(s,s')}}{R(s,s')} ds' = \Omega(s) J(s) + \int_{-L}^{L} \frac{J(s') e^{-ikR(s,s')} - J(s)}{R(s,s')} ds'. \tag{2.2}$$

Here

$$\Omega(s) = \int_{-L}^{L} \frac{ds'}{\sqrt{(s-s')^2 + r^2}} = \Omega + \gamma(s), \tag{2.3}$$

and

$$\gamma(s) = \ln \frac{\left[(L+s) + \sqrt{(L+s)^2 + r^2}\right]\left[(L-s) + \sqrt{(L-s)^2 + r^2}\right]}{4L^2}$$

is a function equal to zero in the vibrator center which attains its largest value on the vibrator's ends, where the current equals zero. In view of boundary conditions

M.V. Nesterenko et al., *Thin Impedance Vibrators*, Lecture Notes in Electrical Engineering 2064, DOI 10.1007/978-1-4419-7850-9_2,

(1.18), $\Omega = 2\ln(2L/r)$ is a large parameter. Then, with (2.3), (2.1) is transformed into the following integrodifferential equation:

$$\frac{d^2J(s)}{ds^2} + k^2J(s) = \alpha\{i\omega E_{0s}(s) + F[s,J(s)] - i\omega z_i J(s)\}, \tag{2.4}$$

where $\alpha = 1/(2\ln[r/(2L)])$ is a natural small parameter ($|\alpha| \ll 1$) and

$$F[s,J(s)] = -\frac{dJ(s')}{ds'}\frac{e^{-ikR(s,s')}}{R(s,s')}\bigg|_{-L}^{L} + \left[\frac{d^2J(s)}{ds^2} + k^2J(s)\right]\gamma(s) \\ + \int_{-L}^{L} \frac{\left[\frac{d^2J(s')}{ds'^2} + k^2J(s')\right]e^{-ikR(s,s')} - \left[\frac{d^2J(s)}{ds^2} + k^2J(s)\right]}{R(s,s')}ds' \tag{2.5}$$

is the vibrator's self-field in free space.

Let us apply the asymptotic averaging method outlined in Sect. 1.5 to obtain the approximate analytical solution of (2.4). To reduce (2.4) to the standard form (1.62) with small parameter in accordance with the method of arbitrary constants variation, we change variables and get

$$\begin{aligned} &J(s) = A(s)\cos ks + B(s)\sin ks, \\ &\frac{dJ(s)}{ds} = -A(s)k\sin ks + B(s)k\cos ks, \quad \left(\frac{dA(s)}{ds}\cos ks + \frac{dB(s)}{ds}\sin ks = 0\right), \\ &\frac{d^2J(s)}{ds^2} + k^2J(s) = -\frac{dA(s)}{ds}\sin ks + \frac{dB(s)}{ds}\cos ks \end{aligned} \tag{2.6}$$

where $A(s)$ and $B(s)$ are the new unknown functions. Then (2.4) is converted into the following system of the integrodifferential equations:

$$\begin{aligned} \frac{dA(s)}{ds} &= -\frac{\alpha}{k}\left\{\begin{matrix} i\omega E_{0s}(s) + F\left[s, A(s), \frac{dA(s)}{ds}, B(s), \frac{dB(s)}{ds}\right] \\ -i\omega z_i[A(s)\cos ks + B(s)\sin ks] \end{matrix}\right\}\sin ks, \\ \frac{dB(s)}{ds} &= +\frac{\alpha}{k}\left\{\begin{matrix} i\omega E_{0s}(s) + F\left[s, A(s), \frac{dA(s)}{ds}, B(s), \frac{dB(s)}{ds}\right] \\ -i\omega z_i[A(s)\cos ks + B(s)\sin ks] \end{matrix}\right\}\cos ks. \end{aligned} \tag{2.7}$$

The obtained equations are equivalent to (2.4) and represent the standard system of integrodifferential equations (1.62), unresolved relative to the derivative. The right-hand sides in (2.7) are proportional to the small parameter α, so the functions $A(s)$ and $B(s)$ on the right-hand sides of (2.7) are slowly changing functions, and the averaging asymptotic method can be used for its solution. Then putting into correspondence the simplified system (1.63) with $dA(s)/ds = 0$ and $dB(s)/ds = 0$ on the right-hand sides and the system (2.7), after performing partial averaging over s explicitly (here the term "partial" means that the averaging operator (1.54) acts on all summands except those containing $E_{0s}(s)$, which is possible ([22] in Chap. 1) for system (2.7)), we obtain the equations of first approximation

$$\begin{aligned}\frac{d\bar{A}(s)}{ds} &= -\alpha\left\{\frac{i\omega}{k}E_{0s}(s) + \bar{F}[s,\bar{A}(s),\bar{B}(s)]\right\}\sin ks + \chi\bar{B}(s),\\ \frac{d\bar{B}(s)}{ds} &= +\alpha\left\{\frac{i\omega}{k}E_{0s}(s) + \bar{F}[s,\bar{A}(s),\bar{B}(s)]\right\}\cos ks - \chi\bar{A}(s),\end{aligned} \tag{2.8}$$

where $\chi = \alpha(i\omega/2k)z_i$, and

$$\bar{F}[s,\bar{A}(s),\bar{B}(s)] = [\bar{A}(s')\sin ks' - \bar{B}(s')\cos ks']\frac{e^{-ikR(s,s')}}{R(s,s')}\bigg|_{-L}^{L} \tag{2.9}$$

is the self-field of the vibrator (2.5), averaged along its length.

We shall obtain the solution of the system (2.8) in the form [1]

$$\begin{aligned}\bar{A}(s) &= C_1(s)\cos\chi s + C_2(s)\sin\chi s,\\ \bar{B}(s) &= -C_1(s)\sin\chi s + C_2(s)\cos\chi s,\end{aligned} \tag{2.10}$$

transforming (2.8) into

$$\begin{aligned}\frac{dC_1(s)}{ds} &= -\alpha\left\{\frac{i\omega}{k}E_{0s}(s) + \bar{F}[s,C_1,C_2]\right\}\sin(k+\chi)s,\\ \frac{dC_2(s)}{ds} &= +\alpha\left\{\frac{i\omega}{k}E_{0s}(s) + \bar{F}[s,C_1,C_2]\right\}\cos(k+\chi)s.\end{aligned} \tag{2.11}$$

Then we obtain $C_1(s)$, $C_2(s)$ from (2.11), and $\bar{A}(s)$, $\bar{B}(s)$ from (2.10), and use these functions as the approximating functions for the current in (2.6). As a result, we obtain the most general asymptotic expression in the parameter α for the current in the thin impedance vibrator with arbitrary excitation:

$$\begin{aligned}J(s) &= \bar{A}(-L)\cos(\tilde{k}s + \chi L) + \bar{B}(-L)\sin(\tilde{k}s + \chi L)\\ &\quad + \alpha\int_{-L}^{s}\left\{\frac{i\omega}{k}E_{0s}(s') + \bar{F}[s',\bar{A},\bar{B}]\right\}\sin\tilde{k}(s-s')ds',\end{aligned} \tag{2.12}$$

where $\tilde{k} = k + \chi = k + i(\alpha/r)\overline{Z}_S$.

To find the constants $\bar{A}(\pm L)$ and $\bar{B}(\pm L)$, it is necessary to use the boundary conditions (1.18) and the conditions of symmetry ([11] in Chap. 1) related to the method of vibrator excitation. If $E_{0s}(s) = E^{s}_{0s}(s)$, then $J(s) = J(-s) = J^{s}(s)$ and $\bar{A}(-L) = \bar{A}(+L)$, $\bar{B}(-L) = -\bar{B}(+L)$; if $E_{0s}(s) = E^{a}_{0s}(s)$, then $J(s) = -J(-s) = J^{a}(s)$ and $\bar{A}(-L) = -\bar{A}(+L)$, $\overline{B}(-L) = \overline{B}(+L)$. Then for symmetric (index "s") and antisymmetric (index "a") current components, we finally obtain, for arbitrary excitation $E_{0s}(s) = E^{s}_{0s}(s) + E^{a}_{0s}(s)$,

$$J(s) = J^{s}(s) + J^{a}(s) = \alpha\frac{i\omega}{k}\left\{\int_{-L}^{s} E_{0s}(s')\sin\tilde{k}(s-s')ds' - \frac{\sin\tilde{k}(L+s) + \alpha P^{s}[kr,\tilde{k}(L+s)]}{\sin 2\tilde{k}L + \alpha P^{s}(kr,2\tilde{k}L)}\int_{-L}^{L} E^{s}_{0s}(s')\sin\tilde{k}(L-s')ds' - \frac{\sin\tilde{k}(L+s) + \alpha P^{a}[kr,\tilde{k}(L+s)]}{\sin 2\tilde{k}L + \alpha P^{a}(kr,2\tilde{k}L)}\int_{-L}^{L} E^{a}_{0s}(s')\sin\tilde{k}(L-s')ds'\right\}, \tag{2.13}$$

where P^{s} and P^{a} are the vibrator self-field functions, given by

$$P^{s}[kr,\tilde{k}(L+s)] = \int_{-L}^{s}\left[\frac{e^{-ikR(s',-L)}}{R(s',-L)} + \frac{e^{-ikR(s',L)}}{R(s',L)}\right]\sin\tilde{k}(s-s')ds'\Bigg|_{s=L} = P^{s}(kr,2\tilde{k}L), \tag{2.14a}$$

$$P^{a}[kr,\tilde{k}(L+s)] = \int_{-L}^{s}\left[\frac{e^{-ikR(s',-L)}}{R(s',-L)} - \frac{e^{-ikR(s',L)}}{R(s',L)}\right]\sin\tilde{k}(s-s')ds'\Bigg|_{s=L} = P^{a}(kr,2\tilde{k}L). \tag{2.14b}$$

2.2 Vibrator Excitation in the Center by Concentrated EMF

To validate the accuracy and to find the limits of applicability of (2.13), we shall discuss the classical problem of vibrator excitation in the geometrical center by lump EMF with amplitude V_0. The mathematical model of excitation may be represented as

$$E_{0s}(s) = E^{s}_{0s}(s) = V_0\delta(s-0), \tag{2.15}$$

where $\delta(s-0) = \delta(s)$ is Dirac's delta function. Then the expression for the current has the form

$$J(s) = -\alpha V_0 \left(\frac{\mathrm{i}\omega}{2\tilde{k}}\right) \frac{\sin \tilde{k}(L-|s|) + \alpha P^{\mathrm{s}}_{\delta}(kr, \tilde{k}s)}{\cos \tilde{k}L + \alpha P^{\mathrm{s}}_{\mathrm{L}}(kr, \tilde{k}L)}. \tag{2.16}$$

Here $P^{\mathrm{s}}_{\delta}(kr, \tilde{k}s) = P^{\mathrm{s}}[kr, \tilde{k}(L+s)] - (\sin \tilde{k}s + \sin \tilde{k}|s|)P^{\mathrm{s}}_{\mathrm{L}}(kr, \tilde{k}L), P^{\mathrm{s}}[kr, \tilde{k}(L+s)]$ is defined by (2.14a), and $P^{\mathrm{s}}_{\mathrm{L}}(kr, \tilde{k}L) = \int\limits_{-L}^{L} (\mathrm{e}^{-\mathrm{i}kR(s,L)}/R(s,L)) \cos \tilde{k}s\, \mathrm{d}s$.

It is possible to obtain $P^{\mathrm{s}}_{\delta}(kr, \tilde{k}s)$ and $P^{\mathrm{s}}_{\mathrm{L}}(kr, \tilde{k}L)$ in explicit form by the technique of generalized integral functions (see Appendix C). Here we give an expression for $P^{\mathrm{s}}_{\mathrm{L}}(kr, \tilde{k}L)$:

$$\begin{aligned} P^{\mathrm{s}}_{\mathrm{L}}(kr, \tilde{k}L) = {} & \cos \tilde{k}L \{2 \ln 2 - \gamma(L) - \frac{1}{2}[\mathrm{Cin}(2\tilde{k}L + 2kL) + \mathrm{Cin}(2\tilde{k}L - 2kL)] \\ & - \frac{\mathrm{i}}{2}[\mathrm{Si}(2\tilde{k}L + 2kL) - \mathrm{Si}(2\tilde{k}L - 2kL)]\} \\ & + \sin \tilde{k}L \left\{ \frac{1}{2}[\mathrm{Si}(2\tilde{k}L + 2kL) + \mathrm{Si}(2\tilde{k}L - 2kL)] \right. \\ & \left. - \frac{\mathrm{i}}{2}[\mathrm{Cin}(2\tilde{k}L + 2kL) - \mathrm{Cin}(2\tilde{k}L - 2kL)] \right\}, \end{aligned} \tag{2.17}$$

where Si(x) and Cin(x) are the integral sine and cosine of the complex argument.

Expression (2.16), in contrast to the solution of the integrodifferential equation (2.4) for the vibrator current by the small parameter method, is given in [2] as

(a) For a tuned vibrator ($\tilde{\tilde{k}}L = n(\pi/2)$, where n is the integer),

$$J_0(s) = C_1 \cos \tilde{\tilde{k}}s + C_2 \sin \tilde{\tilde{k}}s, \tag{2.18a}$$

(b) For an untuned vibrator ($\tilde{\tilde{k}}L \neq n(\pi/2)$),

$$\begin{aligned} J(s) = \alpha J_1(s) = -\alpha \frac{\mathrm{i}\omega/\tilde{\tilde{k}}}{\sin 2\tilde{\tilde{k}}L} & \left\{ \sin \tilde{\tilde{k}}(L-s) \int\limits_{-L}^{s} E_{0\mathrm{s}}(s') \sin \tilde{\tilde{k}}(L+s') \mathrm{d}s' \right. \\ & \left. + \sin \tilde{\tilde{k}}(L+s) \int\limits_{s}^{L} E_{0\mathrm{s}}(s') \sin \tilde{\tilde{k}}(L-s') \mathrm{d}s' \right\}, \end{aligned} \tag{2.18b}$$

or after substituting $E_{0\mathrm{s}}(s) = V_0 \delta(s)$,

$$J(s) = -\alpha V_0 \frac{\mathrm{i}\omega}{2\tilde{\tilde{k}}} \frac{\sin \tilde{\tilde{k}}(L-|s|)}{\cos \tilde{\tilde{k}}L}, \tag{2.18c}$$

where

$$\tilde{\tilde{k}} = k\sqrt{\frac{1+\mathrm{i}\alpha\omega z_i}{k^2}} = k\sqrt{\frac{1+\mathrm{i}2\alpha\bar{Z}_S}{(kr)}}\Bigg|_{|\mathrm{i}2\alpha\bar{Z}_S/(kr)|\angle\angle 1} \approx k + \mathrm{i}\frac{\alpha}{r}\bar{Z}_S = \tilde{k}. \tag{2.19}$$

Solution of (2.4) by the iterations method (see Sect. 1.4.2) for the current in the zeroth and first approximations (with accuracy α^2 inclusive) has the form

$$J_0(s) = -\alpha V_0 \frac{\mathrm{i}\omega}{2k}\frac{\sin k(L-|s|)}{\cos kL}, \tag{2.20a}$$

$$J_1(s) = -\alpha V_0 \frac{\mathrm{i}\omega}{2k}\frac{\sin k(L-|s|) + \alpha F_1(kr, ks, z_i)}{\cos kL + \alpha F(kr, kL, z_i)}. \tag{2.20b}$$

In addition to the above-mentioned solutions, King and Wu have obtained the so-called trinomial formula for the current on a vibrator centrally excited by the δ-generator [3, 4],

$$J(s) = -\alpha_K V_0 \frac{\mathrm{i}\omega}{2\tilde{\tilde{k}}}\frac{\sin\tilde{\tilde{k}}(L-|s|) + F_{K1}(\cos\tilde{\tilde{k}}s - \cos\tilde{\tilde{k}}L) + F_{K2}\left(\cos\frac{ks}{2} - \cos\frac{kL}{2}\right)}{\cos\tilde{\tilde{k}}L}, \tag{2.21}$$

with $\tilde{\tilde{k}}$ defined in (2.19) and

$$\alpha_K = \frac{1}{\Omega_K}, \quad \Omega_K = \begin{cases} \dfrac{|\Omega_K(0)|}{\sin kL}, & kL \leqslant \pi/2, \\ \left|\Omega_K\left(\frac{L-\lambda}{4}\right)\right|, & kL \geqslant \pi/2, \end{cases} \tag{2.22a}$$

$$\Omega_K(s) = \int\limits_{-L}^{L} \frac{e^{-\mathrm{i}kR(s,s')}}{R(s,s')} \sin k(L-|s'|)\mathrm{d}s'. \tag{2.22b}$$

Coefficients F_{K1} and F_{K2}have been found approximately by transforming (2.1) into the Hallen linearized equation (1.42) using its kernel properties. It should be noted that (2.21) coincides with (2.18c) when $F_{K1} = 0$ and $F_{K2} = 0$.

Thus, the solution of integrodifferential equation (2.1) by the averaging method is given by (2.13), valid (in contrast to the solution by the small parameter method) both for tuned ($\sin 2\tilde{k}L = 0$) and untuned ($\sin 2\tilde{k}L \neq 0$) vibrators under arbitrary excitation. The solution of (2.1) by the iterations method requires that the impressed sources field be specified at the initial stage of problem solution. What is more, the distributed vibrator impedance begins to exhibit itself, as follows from (2.20) (in contrast to the solutions by the averaging and the small parameter methods), in the first and succeeding approximations in the small parameter. And finally, the

King–Wu trinomial formula requires different current representations for the tuned and untuned vibrators as in the small parameter method.

The true current distribution (2.16) allows one to calculate the electrodynamic characteristics of the impedance vibrator. Thus, we may obtain the following expression for the vibrator input impedance at the feed point $Z_{in} = R_{in} + iX_{in}$ (or the input admittance $Y_{in} = G_{in} + iB_{in} = 1/Z_{in}$):

$$Z_{in}[\mathrm{ohm}] = \frac{V_0}{J(0)} = \left(\frac{60i\tilde{k}}{\alpha k}\right)\frac{\cos\tilde{k}L + \alpha P^{s}_{L}(kr,\tilde{k}L)}{\sin\tilde{k}L + \alpha P_{\delta L}(kr,\tilde{k}L)}, \tag{2.23}$$

where

$$\begin{aligned} P_{\delta L}(kr,\tilde{k}L) &= \int_{-L}^{L} \frac{e^{-ikR(s,L)}}{R(s,L)} \sin\tilde{k}|s|ds \\ &= \sin\tilde{k}L\Big\{-\gamma(L) + \frac{1}{2}[\mathrm{Cin}(2\tilde{k}L + 2kL) \\ &\quad -\mathrm{Cin}(2\tilde{k}L - 2kL)] - \mathrm{Cin}(\tilde{k}L + kL) + \mathrm{Cin}(\tilde{k}L - kL) \\ &\quad +\frac{i}{2}[\mathrm{Si}(2\tilde{k}L + 2kL) - \mathrm{Si}(2\tilde{k}L - 2kL)] - i[\mathrm{Si}(\tilde{k}L + kL) - \mathrm{Si}(\tilde{k}L - kL)]\Big\} \\ &\quad + \cos\tilde{k}L\Big\{\frac{1}{2}[\mathrm{Si}(2\tilde{k}L + 2kL) + \mathrm{Si}(2\tilde{k}L - 2kL)] - \mathrm{Si}(\tilde{k}L + kL) - \mathrm{Si}(\tilde{k}L - kL) \\ &\quad -\frac{i}{2}[\mathrm{Cin}(2\tilde{k}L + 2kL) + \mathrm{Cin}(2\tilde{k}L - 2kL)] + i[\mathrm{Cin}(\tilde{k}L + kL) + \mathrm{Cin}(\tilde{k}L - kL)]\Big\}. \end{aligned} \tag{2.24}$$

Then the voltage standing wave ratio (VSWR) in the feeder line with characteristic impedance W equals

$$\mathrm{VSWR} = \frac{1 + |S_{11}|}{1 - |S_{11}|}, \quad S_{11} = \frac{Z_{in} - W}{Z_{in} + W}, \tag{2.25}$$

where S_{11} is the reflection coefficient in the feeder.

Let us present some numerical results. Figures 2.1–2.5 show the current amplitude–phase distributions $J(s) = |J(s)|e^{i\arg J(s)}$ in thin ($r/\lambda = 0.007022$) perfectly conducting vibrators with different electrical lengths, calculated with (2.16), in comparison with the experimental data from [5]. As can be seen, the trend of theoretical curves follows that of experimental results quite satisfactorily, with some differences in the absolute values. Such differences are also present in the vibrator input characteristics $Y_{in} = f(2L/\lambda)$ and $|S_{11}| = f(kL)$, calculated by (2.23) and (2.25) and shown in Figs. 2.6 and 2.7. In Fig. 2.7 the theoretical curves corresponding to the King–Middleton solution of Hallen's equation by the iterations method in the second approximation ([11, 12] in Chap. 1) are also plotted, namely

Fig. 2.1 The current amplitude–phase distribution on a perfectly conducting vibrator ($r/\lambda = 0.007022$ and $2L/\lambda = 0.5$): *1* the calculation (2.16); *2* the experimental data [5]

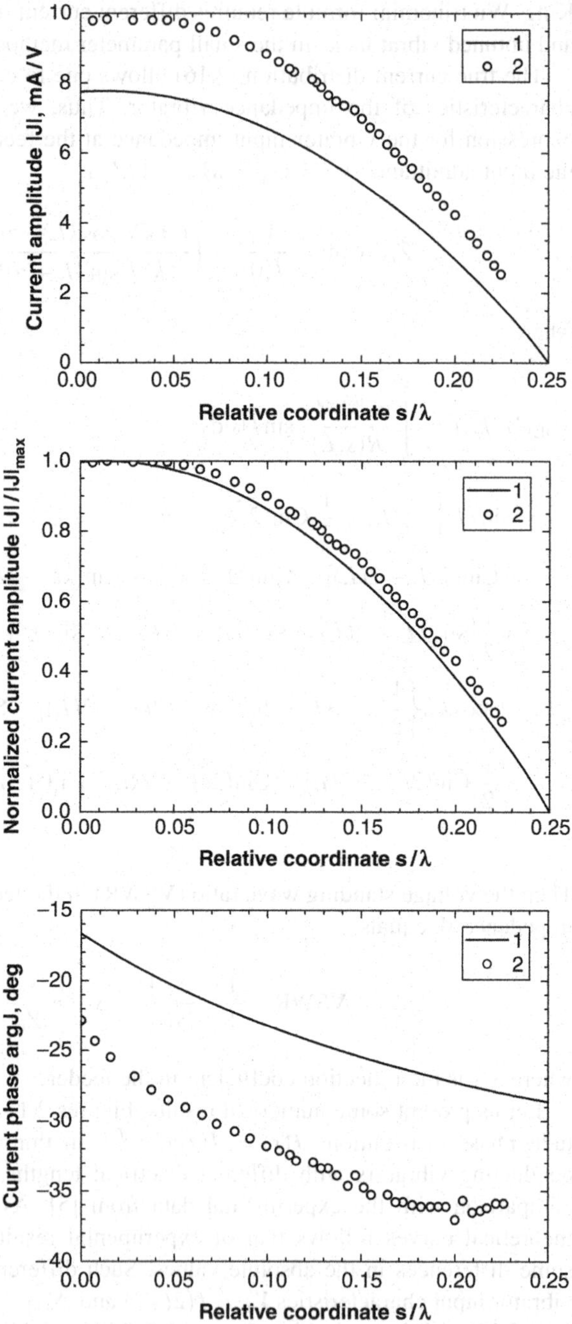

Fig. 2.2 The current amplitude–phase distribution on a perfectly conducting vibrator ($r/\lambda = 0.007022$ and $2L/\lambda = 0.75$): *1* the calculation (2.16); *2* the experimental data [5]

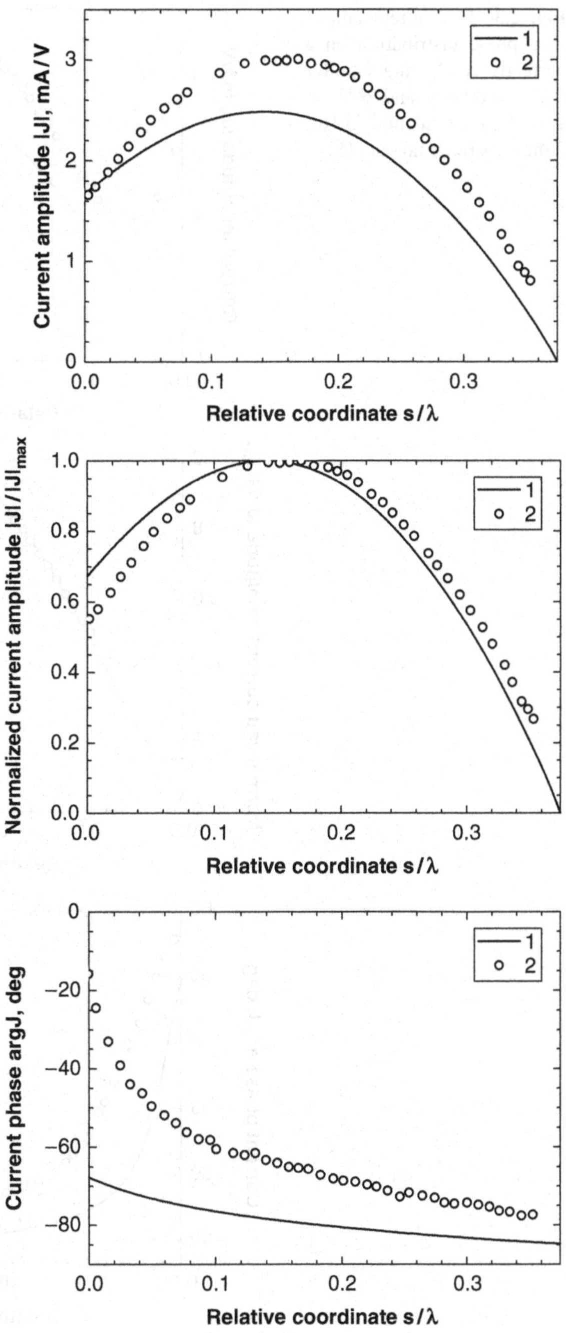

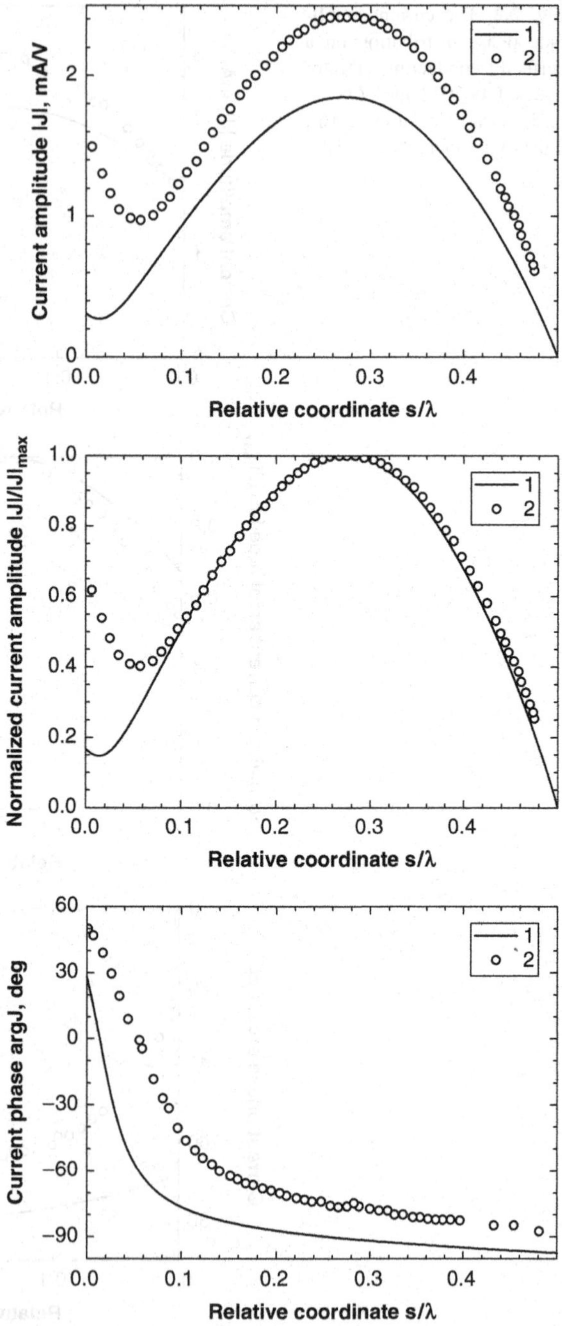

Fig. 2.3 The current amplitude–phase distribution on a perfectly conducting vibrator ($r/\lambda = 0.007022$ and $2L/\lambda = 1.0$): *1* the calculation (2.16); *2* the experimental data [5]

Fig. 2.4 The current amplitude–phase distribution on a perfectly conducting vibrator ($r/\lambda = 0.007022$ and $2L/\lambda = 1.25$): *1* the calculation (2.16); *2* the experimental data [5]

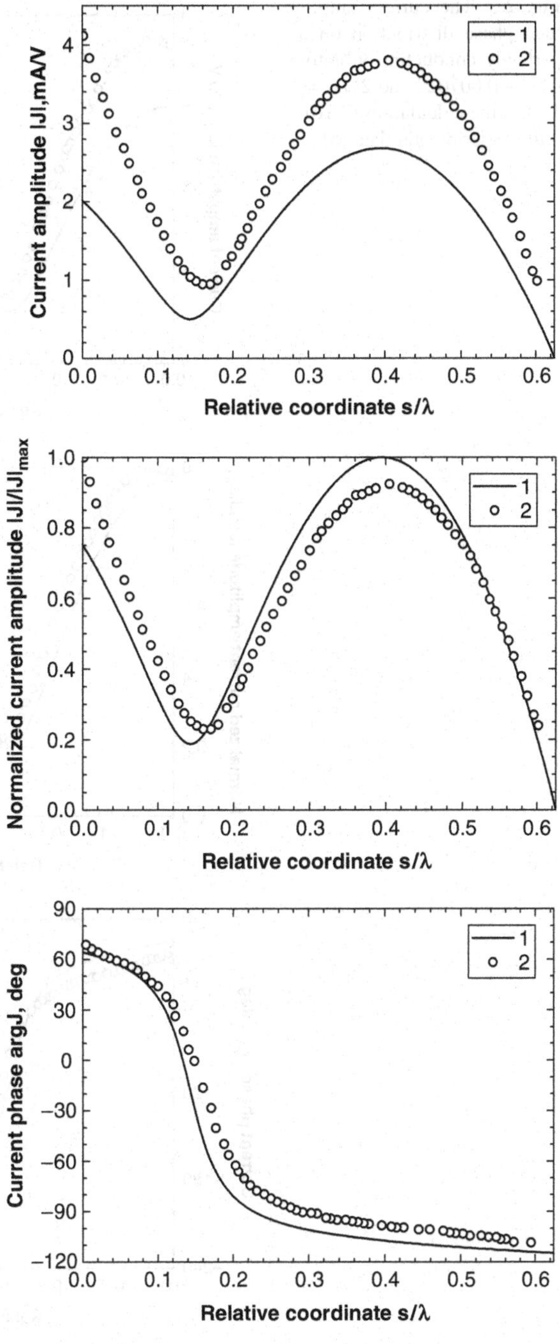

Fig. 2.5 The current amplitude–phase distribution on a perfectly conducting vibrator ($r/\lambda = 0.007022$ and $2L/\lambda = 1.5$): *1* the calculation (2.16); *2* the experimental data [5]

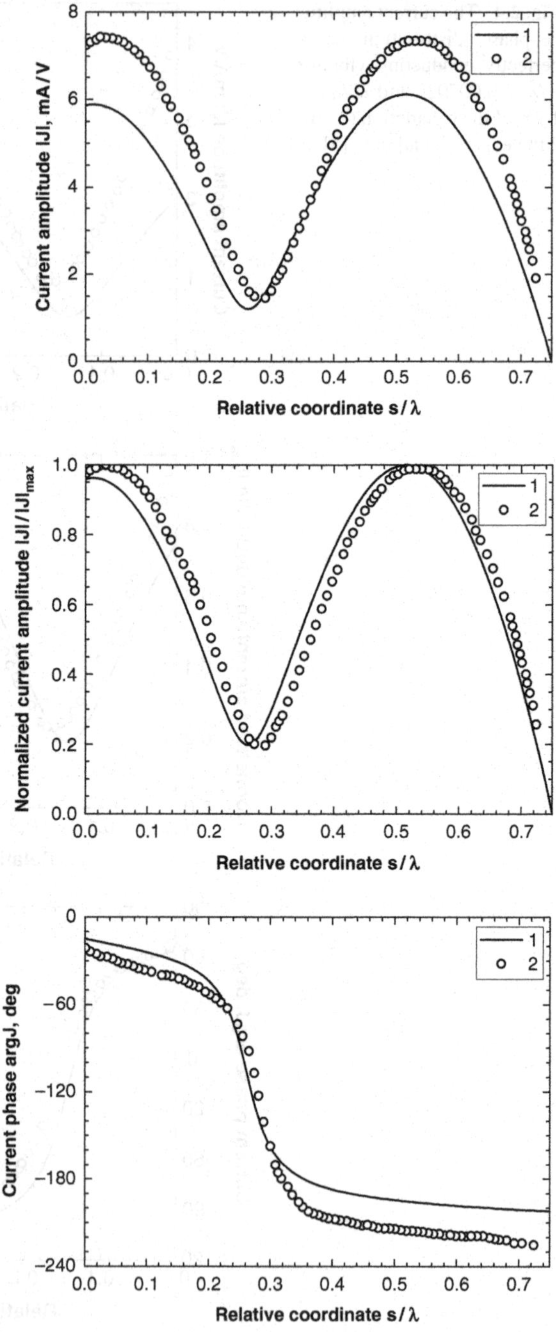

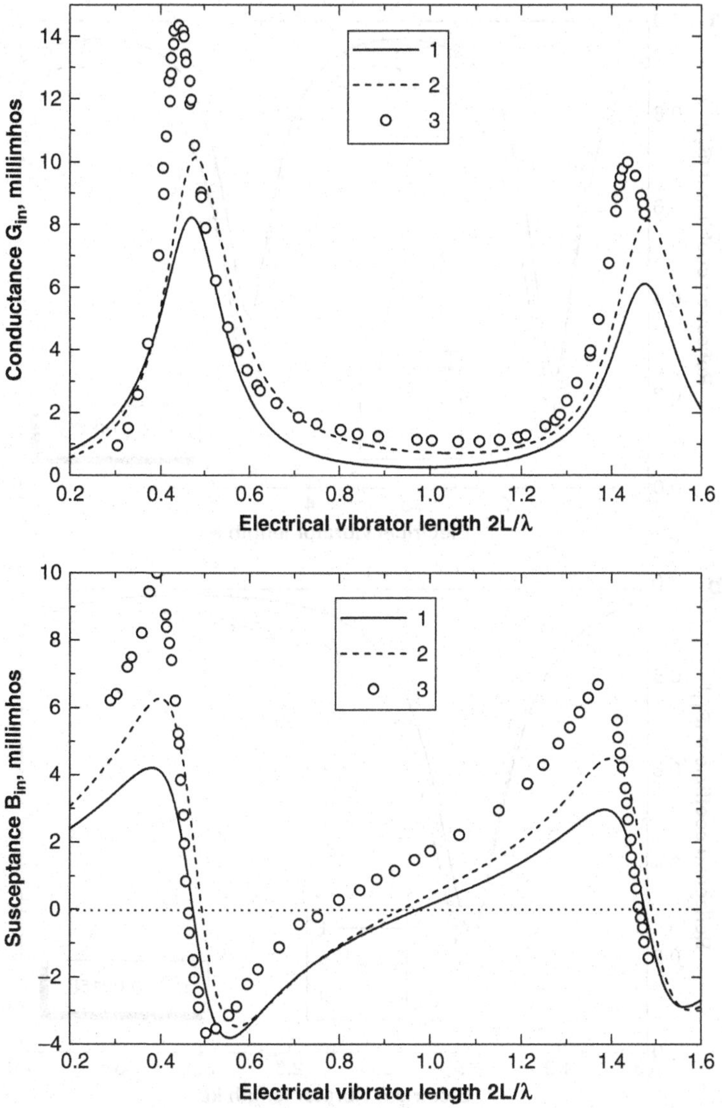

Fig. 2.6 The input admittance of a perfectly conducting vibrator versus electrical length ($r/\lambda = 0.007022$): *1* the calculation (2.23); *2* the calculation (2.27); *3* the experimental data [5]

$$Y^K_{\text{in}\,2}[\text{millimhos}] = \frac{i\alpha_K}{60}\,\frac{\sin kL + \alpha_K F_{1s}(kr, kL) + \alpha_K^2 F_{2s}(kr, kL)}{\cos kL + \alpha_K F_1(kr, kL) + \alpha_K^2 F_2(kr, kL)}, \tag{2.26}$$

where α_K is defined by (2.22).

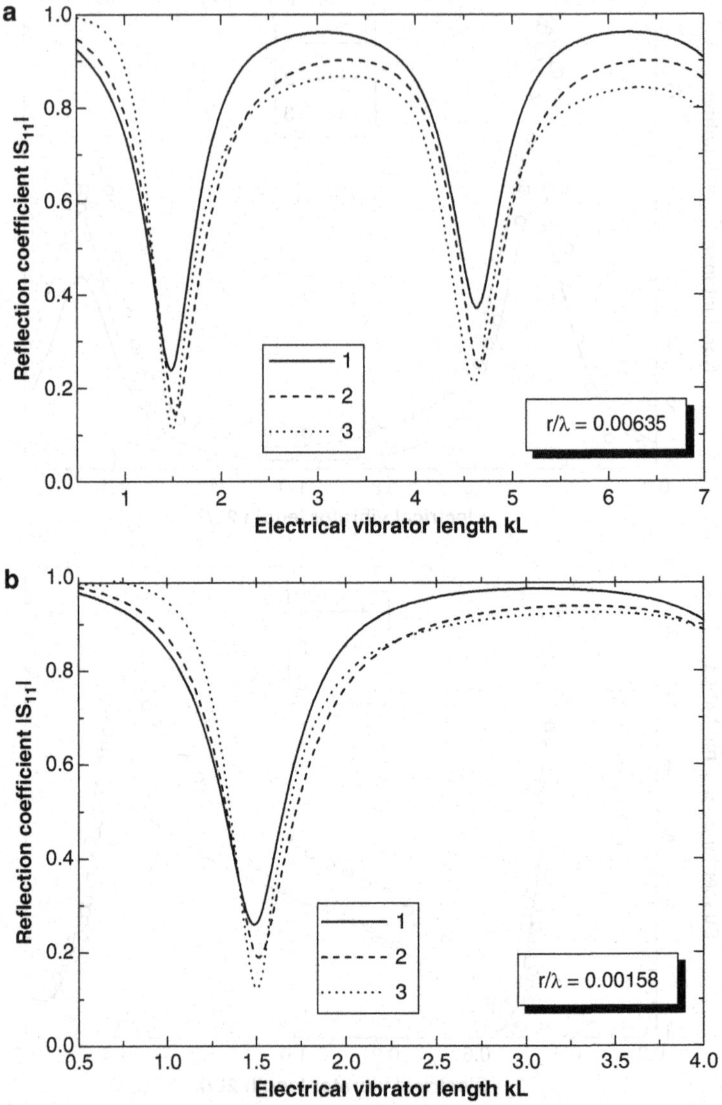

Fig. 2.7 The reflection coefficient in the feeder with $W = 75$ ohm versus electrical length of a perfectly conducting vibrator: *1* the calculation (2.23); *2* the calculation (2.27); *3* the calculation (2.26)

An analogous situation is observed for the input characteristics of impedance vibrators. The plots of input admittance for two different surface impedances are represented in Figs. 2.8 and 2.9: (1) a metallic conductor (radius $r_i = 0.3175$ cm)

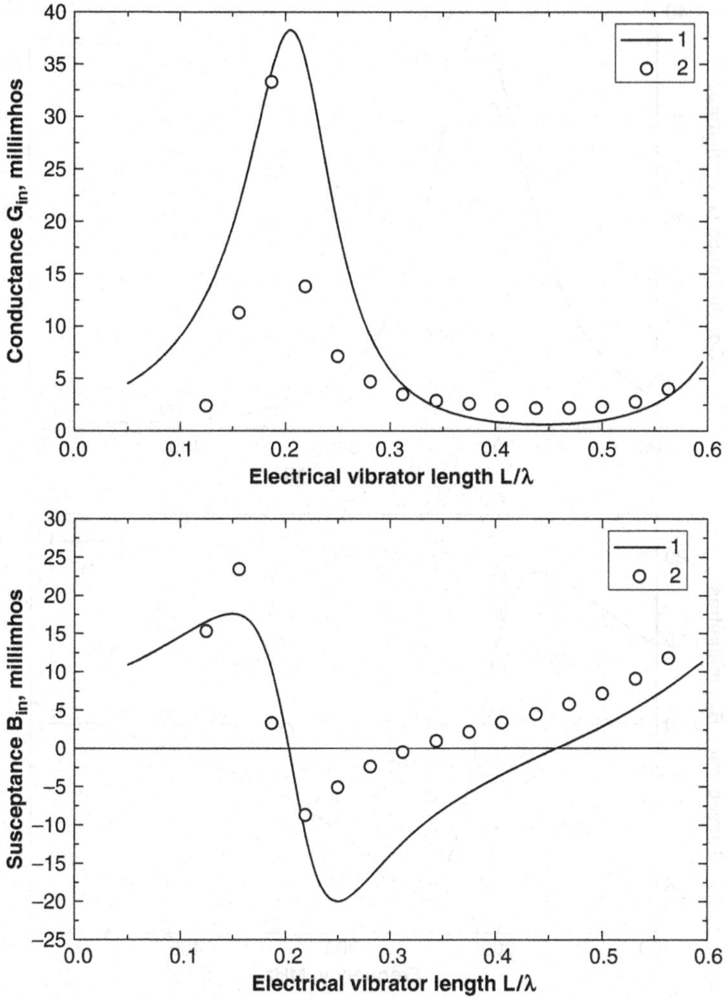

Fig. 2.8 Input admittance of a metallic conductor of radius $r_i = 0.3175\,\text{cm}$, covered by a dielectric ($\varepsilon = 9.0$) shell with radius $r = 0.635\,\text{cm}$ versus electrical length at 600 MHz: *1* the calculation (2.23); *2* the experimental data [6]

covered by a dielectric ($\varepsilon = 9.0$) shell (radius $r = 0.635\,\text{cm}$), where Fig. 2.8 shows experimental data from [6]; (2) a metallic conductor (radius $r_i = 0.5175\,\text{cm}$) covered by a ferrite ($\mu = 4.7$) shell (radius $r = 0.6\,\text{cm}$), where Fig. 2.9 shows experimental data from [7].

Differences among the theoretical curves obtained from solution of the integral equation by the averaging method, the experimental data, and the graphs plotted by

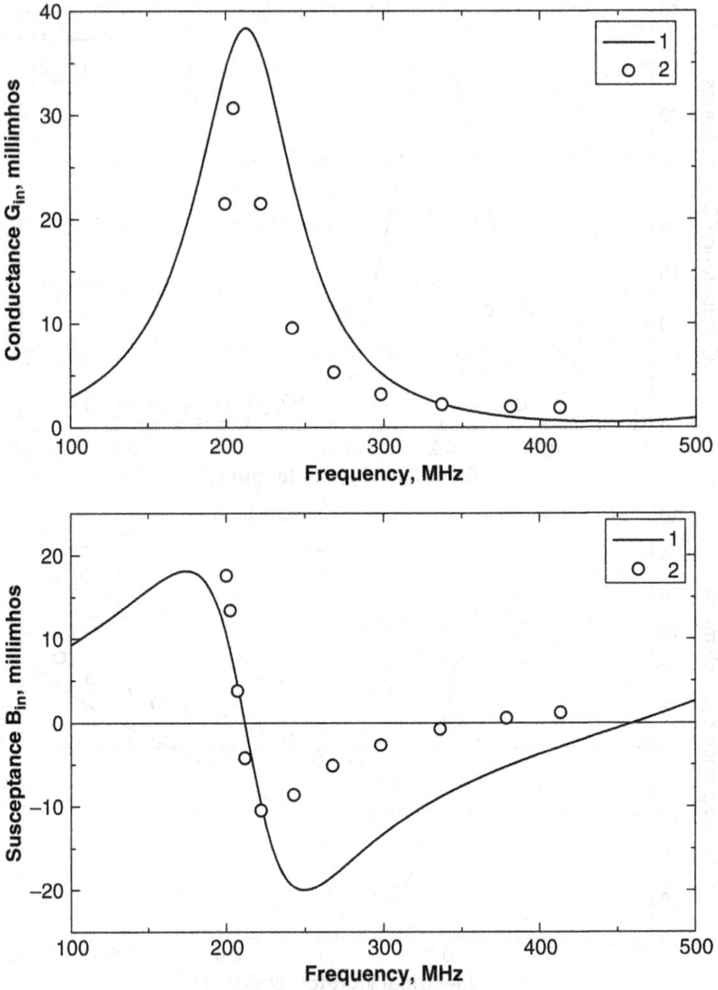

Fig. 2.9 Input admittance of a metallic conductor of radius $r_i = 0.5175$ cm covered by a ferrite ($\mu = 4.7$) shell with radius $r = 0.6$ cm versus frequency at $L = 30$ cm: *1* the calculation (2.23); *2* the experimental data [7]

higher approximations of the iterations method may be explained by errors in averaging of the self-field of the vibrator (2.5). However, the resonant characteristics of the vibrators $\left((2L/\lambda)_{\text{res}}\right.$ for $B_{\text{in}} = 0$ and $(kL)_{\text{res}}$ for $\left.|S_{11}| = |S_{11}|_{\min}\right)$ are defined rather precisely, and the calculated curves for the normalized current amplitudes $\left(|J(s)|/|J|_{\max}\right)$ agree with the experimental data within acceptable limits. Thus, the formulas for the current obtained by a first approximation of the averaging method are applicable to the calculation of vibrator integral characteristics such as the radiated (scattered) electromagnetic field in all field zones, and to the investigation of the resonant properties of the vibrator.

As was indicated in Sect. 1.5, the solution of (2.7) can be obtained by an improved first approximation. This means that the transition from (2.7) to (2.8) is accomplished by the substitution $-(\mathrm{d}\bar{A}(s)/\mathrm{d}s)\sin ks + (\mathrm{d}\bar{B}(s)/\mathrm{d}s)\cos ks = \alpha \mathrm{i}\omega E_{0s}(s)$. Then the input impedance of the vibrator is

$$Z_{\text{in}}^{\text{imp}} = \left(\frac{60\mathrm{i}\tilde{k}}{\alpha k}\right) \frac{\cos\tilde{k}L + \alpha P_{\text{L}}^{\text{s}}(kr,\tilde{k}L)}{\sin\tilde{k}L + \alpha P_{\delta L}(kr,\tilde{k}L) + [\sin\tilde{k}L + \alpha P_{\delta 1}(kr,\tilde{k}L) + \alpha^2 P_{\delta 2}(kr,\tilde{k}L)]}. \tag{2.27}$$

Here

$$\begin{aligned}
&P_{\delta 1}(kr,\tilde{k}L) = P_{\delta \text{L}}(kr,\tilde{k}L) + \sin\tilde{k}LP_0^{\text{s}}(kr,\tilde{k}L) - \cos\tilde{k}LP_{\delta 0}(kr,\tilde{k}L),\\
&P_{\delta 2}(kr,\tilde{k}L) = P_{\delta \text{L}}(kr,\tilde{k}L)P_0^{\text{s}}(kr,\tilde{k}L) - P_{\text{L}}^{\text{s}}(kr,\tilde{k}L)P_{\delta 0}(kr,\tilde{k}L),\\
&P_0^{\text{s}}(kr,\tilde{k}L) = \int_{-L}^{L} \frac{\mathrm{e}^{-\mathrm{i}kR(s,0)}}{R(s,0)}\cos\tilde{k}s\,\mathrm{d}s, \quad P_{\delta 0}(kr,\tilde{k}L) = \int_{-L}^{L} \frac{\mathrm{e}^{-\mathrm{i}kR(s,0)}}{R(s,0)}\sin\tilde{k}|s|\mathrm{d}s.
\end{aligned} \tag{2.28}$$

The curves calculated by (2.27) are given in Figs. 2.6 and 2.7 as dotted lines, well correlated with the solution (2.26). The cumbersome formulas derived by solving (2.7) by the averaging method to second approximation improve the accuracy of the results but are useless in practice. However, the accuracy of a mathematical model may be enhanced by other methods, as will be demonstrated below.

2.2.1 *Impedance Vibrator with Lumped Load in the Center*

The problem of impedance vibrator excitation by an EMF δ-generator can be used for analysis of passive vibrators with lumped load. The current in a symmetric vibrator loaded by lumped impedance Z_{cL} at $s = 0$ and located in the field of a plane electromagnetic wave is defined by a combination of two current distributions ([11] in Chap. 1):

$$J_r(s) = J_{\text{sc}}(s) - J_{\text{tr}}(s), \tag{2.29}$$

where $J_{\text{sc}}(s)$ is the current in the gapless vibrator and $J_{\text{tr}}(s)$ is the current in the vibrator excited by the δ-generator. The current $J_{\text{sc}}(s)$ in the gapless scattering vibrator with accuracy to the terms of order α^2 is given, according to (2.13), by

$$J_{sc}(s) = -\alpha E_0 \cos\psi \sin\theta \frac{i\omega/(k\tilde{k})}{1-(q/\tilde{k})^2}$$
$$\times \left[\frac{\cos\tilde{k}s \cos qL - \cos\tilde{k}L \cos qs}{\cos\tilde{k}L + \alpha P_L^s(kr,\tilde{k}L)} + i\frac{\sin\tilde{k}s \sin qL - \sin\tilde{k}L \sin qs}{\sin\tilde{k}L + \alpha P_L^a(kr,\tilde{k}L)}\right], \tag{2.30}$$

where $E_{0s}(s) = E_0 \cos\psi \sin\theta\, e^{iks\cos\theta}$. Here $q = k\cos\theta$, E_0 is the incident wave amplitude, ψ is the angle between the vibrator axis and the polarization plane of the incident wave, the angle θ is measured from the vibrator axis, and

$$P_L^a(kr,\tilde{k}L) = -\int_{-L}^{L} G(s,L)\sin\tilde{k}s\,ds$$
$$= \sin\tilde{k}L\left\{2\ln 2 - \gamma(L) - \frac{1}{2}[\mathrm{Cin}(2\tilde{k}L+2kL)+\mathrm{Cin}(2\tilde{k}L-2kL)]\right.$$
$$\left. - \frac{i}{2}[\mathrm{Si}(2\tilde{k}L+2kL)-\mathrm{Si}(2\tilde{k}L-2kL)]\right\}$$
$$-\cos\tilde{k}L\left\{\frac{1}{2}[\mathrm{Si}(2\tilde{k}L+2kL)+\mathrm{Si}(2\tilde{k}L-2kL)]\right.$$
$$\left. - \frac{i}{2}[\mathrm{Cin}(2\tilde{k}L+2kL)-\mathrm{Cin}(2\tilde{k}L-2kL)]\right\}. \tag{2.31}$$

The current $J_{tr}(s)$ is calculated by (2.16) if V_0 is replaced by V_{cL},

$$V_{cL} = J_{sc}(0)\frac{Z_{in}Z_{cL}}{Z_{in}+Z_{cL}}. \tag{2.32}$$

Then the current in the load of the receiving antenna has the form

$$J_r(0) = J_{sc}(0)\frac{Z_{in}}{Z_{in}+Z_{cL}} = \alpha E_0 \cos\psi \sin\theta \frac{i\omega/(k\tilde{k})}{1-(q/\tilde{k})^2}$$
$$\times \left[\frac{(\cos\tilde{k}L - \cos qL) + \alpha P_L^s(kr,\tilde{k}L)}{\cos\tilde{k}L + \alpha P_L^s(kr,\tilde{k}L)}\right]\frac{Z_{in}}{Z_{in}+Z_{cL}}. \tag{2.33}$$

2.2.2 *Surface Impedance of Thin Vibrators*

As discussed above, in comparative numerical calculations for perfectly conducting vibrators the value of the distributed surface impedance $\overline{Z}_S$ is set equal to zero.

However, the analysis of certain vibrators requires formulas for the numerical estimation of the surface impedance. Let us consider the problem of axisymmetric excitation of an infinite double-layer cylinder with outer radius r and inner radius r_i by a converging cylindrical wave. Let us introduce the cylindrical coordinate system ρ, φ, z with z-axis directed along the cylinder's axis. By symmetry, the electromagnetic field has only E_z and H_φ components, depending only on ρ. The medium has permittivity ε and permeability μ in the region $r - r_i$, and ε_i, μ_i when $\rho \leqslant r_i$.

The surface impedance $\overline{Z}_S = E_z/H_\varphi$ at $\rho = r$ may be found as a solution of Maxwell's equations expressed in terms of Bessel function $I_{0,1}$ and Neumann function $N_{0,1}$ as

$$\begin{Bmatrix} iE_z \\ H_\varphi \end{Bmatrix} = \begin{Bmatrix} I_0(k\sqrt{\varepsilon\mu}r) + N_0(k\sqrt{\varepsilon\mu}r) \\ I_1(k\sqrt{\varepsilon\mu}r) + N_1(k\sqrt{\varepsilon\mu}r) \end{Bmatrix}$$

$$\times\sqrt{\frac{\mu}{\varepsilon}}\frac{\sqrt{\frac{\varepsilon}{\mu}}N_1(k\sqrt{\varepsilon\mu}r_i)I_0(k\sqrt{\varepsilon_i\mu_i}r_i) - \sqrt{\frac{\varepsilon_i}{\mu_i}}N_0(k\sqrt{\varepsilon\mu}r_i)I_1(k\sqrt{\varepsilon_i\mu_i}r_i)}{\sqrt{\frac{\varepsilon_i}{\mu_i}}I_0(k\sqrt{\varepsilon\mu}r_i)I_1(k\sqrt{\varepsilon_i\mu_i}r_i) - \sqrt{\frac{\varepsilon}{\mu}}I_1(k\sqrt{\varepsilon\mu}r_i)I_0(k\sqrt{\varepsilon_i\mu_i}r_i)}. \tag{2.34}$$

Assuming $r_i = 0$ and $|\varepsilon| \gg 1$ ($\varepsilon = \varepsilon' + 4\pi\sigma/i\omega$), we obtain the familiar formula for the impedance of a cylindrical conductor [3, 4], with the skin effect

$$\overline{Z}_S = \frac{k'}{120\pi\sigma}\frac{I_0(k'r)}{I_1(k'r)}, \tag{2.35}$$

where $k' = (1-i)/\Delta^0, \Delta^0 = \omega/k\sqrt{2\pi\sigma\omega\mu}$ is the skin-layer thickness, and σ is the conductivity of the metal.

Consider corrugated ($L_1 \approx L_2$) or ribbed ($L_1 \ll L_2$) conductors (here L_1 is the ridge thickness, where $\overline{Z}_S = 0$, and L_2 is the cavity width, where $\overline{Z}_S \neq 0$) with cell periods $(L_1 + L_2) \ll \lambda/\sqrt{\varepsilon\mu}$ and $|\varepsilon_i| \gg 1$. Averaging the impedances over the cell period and taking into account (2.34), we have

$$\overline{Z}_S = -i\frac{L_2}{L_1 + L_2}\sqrt{\frac{\mu}{\varepsilon}}\frac{I_0(k\sqrt{\varepsilon\mu}r)N_0(k\sqrt{\varepsilon\mu}r_i) - I_0(k\sqrt{\varepsilon\mu}r_i)N_0(k\sqrt{\varepsilon\mu}r)}{I_1(k\sqrt{\varepsilon\mu}r)N_0(k\sqrt{\varepsilon\mu}r_i) - I_0(k\sqrt{\varepsilon\mu}r_i)N_1(k\sqrt{\varepsilon\mu}r)}, \tag{2.36}$$

which is valid for conductors with an isolating covering made of a magneto-dielectric [2] ($L_1 = 0$), and also for metallic cylinders ($r_i = 0$) with transverse dielectric insertions ($L_2 \ll L_1$).

For thin vibrators $\left(\left|(k\sqrt{\varepsilon\mu}r)^2 \ln(k\sqrt{\varepsilon\mu}r_i)\right| \ll 1\right)$, the surface impedance does not depend on the excitation mode, and the corresponding boundary conditions become impressed [8], that is, they do not depend upon the structure of exciting field. Then with (2.34)–(2.36) we find that the complex impedances $\overline{Z}_S = \overline{R}_S + i\overline{X}_S$ for vibrators in thin-wire approximation equal

$$\overline{Z}_S = \frac{1+i}{120\pi\sigma\Delta^0} \tag{2.37}$$

for a solid metallic cylinder if $r \gg \Delta^0$; note that $\overline{Z}_S = 0$ for a perfect conductor $(\sigma \to \infty)$;

$$\overline{Z}_S = \frac{1}{120\pi\sigma h_0 + ikr(\varepsilon - 1)/2} \tag{2.38}$$

for a dielectric cylinder with thin metal covering ($h_0 \ll \Delta^0, \varepsilon = 1$);

$$\overline{Z}_S = \frac{1}{120\pi\sigma h_0} \tag{2.39}$$

for a metallic tubular cylinder $r \ll \Delta^0$ ("nanoradius" vibrator [4] $h_0 = r, r \ll \Delta^0$); and after substitution $h_0 = 0$ in (2.38),

$$\overline{Z}_S = -i\frac{2}{kr(\varepsilon - 1)} \tag{2.40}$$

for a dielectric cylinder;

$$\overline{Z}_S = -i\frac{L_2}{L_1 + L_2}\frac{2}{kr\varepsilon} \tag{2.41}$$

for a metal-dielectric cylinder;

$$\overline{Z}_S = \frac{1}{120\pi\sigma h_0 - i/kr\mu \ln(r/r_i)} \tag{2.42}$$

for a magnetodielectric metalized cylinder with inner conducting cylinder $r = r_i$ (2.39); and if $h_0 = 0$, then

$$\overline{Z}_S = ikr\mu \ln\frac{r}{r_i} \tag{2.43}$$

for a metallic cylinder with magnetodielectric covering (thickness $r - r_i$ [2]) or a ribbed cylinder;

$$\overline{Z}_S = \frac{i}{2}kr\cot^2\psi \tag{2.44}$$

for a metallic monofilar helix with radius $r(kr \ll 1)$ and winding angle ψ.

Formulas (2.37)–(2.44) have been obtained in the framework of impedance conception ([3] in Chap. 1, [8]), and they are valid for thin cylinders both with finite and infinite extension located in free space. If the vibrator is situated in a material medium with parameters ε_1 and μ_1, then all formulas must be multiplied by $\sqrt{\mu_1/\varepsilon_1}$. Since the surface impedance often depends on the parameters ε and μ, it is possible to alter the characteristics of antennas with fixed geometric dimensions by varying these parameters if they depend on the external static electrical and/or magnetic fields. It also follows from (2.43) and (2.44) that it is possible for vibrators with pure inductive surface impedance to define a term known as effective vibrator length $2L_{\text{eff}}$ [9]:

$$2L_{\text{eff}} = \left[1 + \frac{\mu \ln(r/r_i)}{2\ln(2L/r)}\right] 2L, \quad 2L_{\text{eff}} = \left[1 + \frac{\cot^2 \psi}{4\ln(2L/r)}\right] 2L, \tag{2.45}$$

that is, the impedance vibrator length $2L$ is "equivalent" to the perfectly conducting vibrator with the length $2L_{\text{eff}}$ with $2L_{\text{eff}} > 2\text{L}$.

2.2.3 Resonant Properties of Impedance Vibrators in Free Space

Near the resonance, when $\tilde{k}L \approx \pi/2$ and $\sin \tilde{k}L \approx 1$, it is possible to neglect the second summand in the denominator of (2.33), proportional to the small parameter α. If vibrator impedance is purely reactive, i.e., $\overline{R}_S = 0$, then defining the resonant condition as $X_{\text{in}} = 0$, we obtain the transcendental equation, allowing us to find the length (frequency) of the resonant vibrator,

$$\cos(\tilde{k}L)_{\text{res}} + \alpha \operatorname{Re} P_{\text{L}}^{\text{s}}(kr, (\tilde{k}L)_{\text{res}}) = 0, \tag{2.46}$$

where $\operatorname{Re} P_{\text{L}}^{\text{s}}$ is the real part of P_{L}^{s}, defined by (2.17). Let us remark that, as it will be shown below, the real part of the impedance vibrator has no considerable influence on its resonant properties.

Let us obtain an approximate solution of (2.46), expanding the unknown value $(\tilde{k}L)_{\text{res}}$ in a power series in the small parameter α:

$$(\tilde{k}L)_{\text{res}} = (\tilde{k}L)_0 + \alpha(\tilde{k}L)_1 + \alpha^2(\tilde{k}L)_2 + \ldots. \tag{2.47}$$

Substituting (2.47) into (2.46) and equating summands with equal powers, we have

$$(\tilde{k}L)_{\text{res}} \approx \frac{\pi}{2} + \alpha \operatorname{Re} P_{\text{L}}^{\text{s}}\left(\frac{\pi r}{2L}, \frac{\pi}{2}\right), \tag{2.48}$$

with accuracy to terms of order α^2. With (2.17), (2.48) may be transformed into

$$(\tilde{k}L)_{\text{res}} \approx \frac{\pi}{2} + \alpha\left[\frac{1}{2}\text{Si}(2\pi - 2\chi L) + \frac{1}{2}\text{Si}(2\chi L)\right], \tag{2.49}$$

where $\chi = -\alpha(\overline{X}_S/r)$. Assuming $(2L)_{\text{res}} \approx \lambda/2$ and taking into consideration that $\alpha = 1/(2\,\ln(2r/\lambda))$, $2\chi L = -\alpha(\lambda\overline{X}_S/2r)$, $\tilde{k} = k(1-\alpha(\overline{X}_S/kr))$, we obtain the formula for the vibrator resonant length as a function of its radius r, wavelength λ, and surface impedance $\overline{X}_S$:

$$(2L)_{\text{res}} \approx \frac{\lambda}{2}\frac{1}{1-\alpha\frac{\overline{X}_S}{kr}} + \alpha\frac{1}{k\left(1-\alpha\frac{\overline{X}_S}{kr}\right)}\left[\text{Si}\left(2\pi+\alpha\frac{\lambda\overline{X}_S}{2r}\right) - \text{Si}\left(\alpha\frac{\lambda\overline{X}_S}{2r}\right)\right]. \tag{2.50}$$

Expression (2.50) may be simplified for relatively small values of the normalized surface impedance ($\overline{X}_S \sim kr$) and represented as

$$(2L)_{\text{res}} \approx \frac{\lambda}{2} - \frac{\lambda}{4\pi\ln\frac{\lambda}{2r}}\left(\text{Si}(2\pi) + \frac{\lambda\overline{X}_S}{2r}\right). \tag{2.51}$$

We note that (2.51) was derived by the power series expansion

$$\frac{1}{1-\alpha\left(\frac{\overline{X}_S}{kr}\right)} = 1 + \alpha\left(\frac{\overline{X}_S}{kr}\right) - \alpha^2\left(\frac{\overline{X}_S}{kr}\right)^2 + \cdots \approx 1 + \alpha\left(\frac{\overline{X}_S}{kr}\right).$$

It was shown in Sect. 2.2.2 that if $\overline{X}_S>0$ (the inductive impedance has, for example, a metallic conductor covered by a magnetodielectric layer, a corrugated cylindrical conductor, or a monofilar metallic helix), then the surface impedance of a thin vibrator can be represented as $\overline{X}_S = krC_L$; and if $\overline{X}_S < 0$ (capacitive impedance has, for example, a dielectric or a layered metal-dielectric cylinder), then $\overline{X}_S = -kr[C_C/(kr)^2]$, where the constants C_L and C_C are defined by the geometric dimension and electrophysical parameters of the vibrator material. Bearing this in mind, we transform (2.51) into

$$(2L)_{\text{res}} \approx \frac{\lambda}{2} - \frac{\lambda}{4\,\ln\frac{\lambda}{2r}}\left(\frac{\text{Si}(2\pi)}{\pi} + \frac{\overline{X}_S}{kr}\right). \tag{2.52}$$

As follows from (2.52), the resonant length of a thin vibrator in free space can be either shorter or longer than $\lambda/2$ (vibrator shortening or lengthening, respectively), depending on the distributed surface impedance type. Note the principal difference for perfect conductivity ($\overline{Z}_S = 0$), where $(2L)_{\text{res}}<(\lambda/2)$ for any finite r/λ. This is illustrated in Fig. 2.10a, where the normalized resonant length in dependence on the vibrator radius for capacitive and inductive impedances, and also for a perfectly conducting vibrator, is shown. For comparison, results obtained by the iterations method to a second approximation in [12] in Chap. 1 (the formula (2.26)) are presented. For the capacitive impedance (curves 1 and 2), lengthening $(2L)_{\text{res}}$

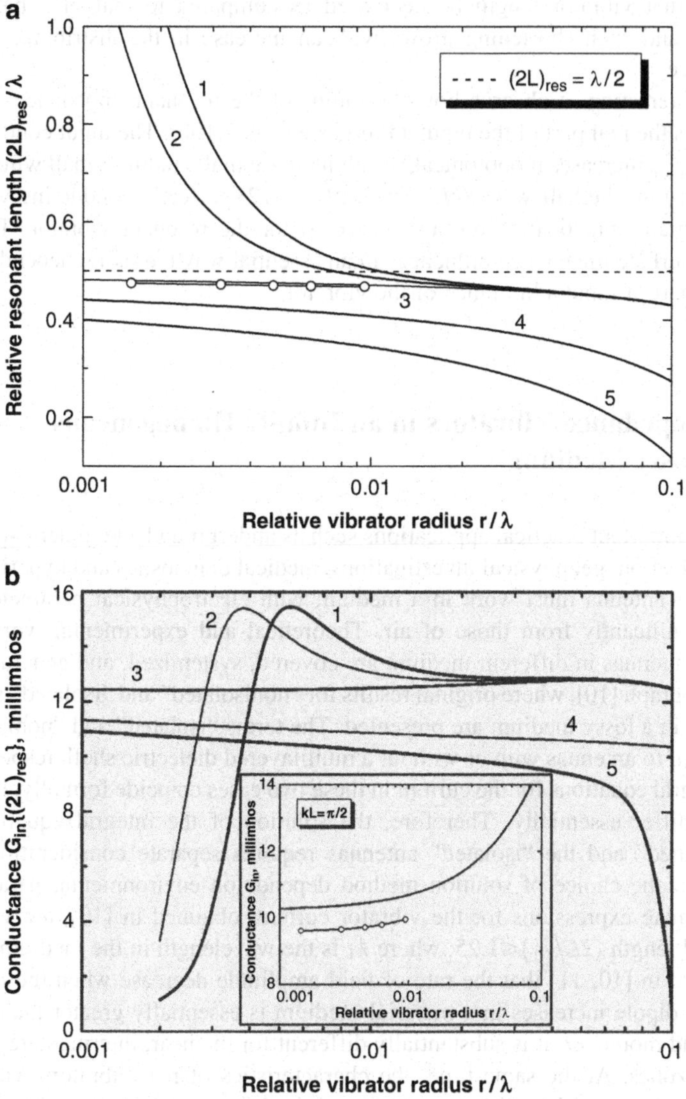

Fig. 2.10 The relative resonant length and the input conductance versus vibrator radius: *1* $\overline{Z}_S = -i\overline{X}_S(C_C = 0.002)$; *2* $\overline{Z}_S = -i\overline{X}_S(C_C = 0.001)$; *3* $\overline{Z}_S = 0$ (the *circles* correspond to calculation [12] in Chap. 1); *4* $\overline{Z}_S = i\overline{X}_S(C_L = 1.0)$; *5* $\overline{Z}_S = i\overline{X}_S(C_L = 2.0)$

transits to shortening as the radius increases, reaching for some r/λ the value $(2L)_{\text{res}} = \lambda/2$ (a half-wave vibrator). For a perfectly conducting vibrator (curve 3 and the circles) and for inductive impedance (curves 4 and 5), the resonant tuning requires that vibrator length be decreased as compared to that of a half-wave vibrator, and such shortening grows with an increase in the distributed surface impedance.

It is interesting to observe how the radius of the resonant impedance vibrator influences the real part of the input admittance (Fig. 2.10b). The input conductance $G_{\text{in}}\{(2L)_{\text{res}}\}$ increases monotonically with increase in the radius (small windows in Fig. 2.10b) for the half-wave ($kL = \pi/2$, $2L = \lambda/2$) perfectly conducting vibrator, but it remains practically constant (curve 3) for the resonant vibrator. The distributed surface impedance influences in an essential way the radial dependence of the real part of input admittance of the vibrator.

2.3 Impedance Vibrators in an Infinite Homogeneous Lossy Medium

In some important practical applications such as underground and underwater radio communication, geophysical investigations, medical diagnostics and hyperthermia, a vibrator antenna must work in a medium with electrophysical parameters that differ significantly from those of air. Theoretical and experimental works concerning antennas in different medium are covered, systemized, and generalized in the monograph [10], where original results for "nonisolated" and "isolated" vibrator antennas in a lossy medium are presented. The terms "isolated" and "nonisolated" are related to antennas with or without a multilayered dielectric shell, respectively. The integral equations for the current in these two cases coincide formally, but their kernels differ essentially. Therefore, the solution of the integral equations for "nonisolated" and the "isolated" antennas requires separate considerations, and moreover, the choice of solution method depends on environmental parameters. Approximate expressions for the vibrator current, obtained in [10], are valid for electrical length $(2L/\lambda_1) \leqslant 1.25$, where λ_1 is the wavelength in the medium. It was also noted in [10, 11] that the rate of field amplitude decrease when the distance from the dipole increases in a material medium is essentially greater than in free space, and moreover, it is substantially different for the near, intermediate, and far antenna zones. At the same time, the characteristics of real vibrators with finite dimensions comparable with the wavelength in the material medium differ essentially from the corresponding parameters of electrically short dipoles. Hence, taking into account possible fields of application of vibrator antennas located in different medium a thorough analysis of the spatial field distribution in a near-field zone of the vibrator (especially when it has complex distributed impedance) is of real practical interest.

In this section we will consider thin impedance vibrators located in an infinite homogeneous medium with sufficiently arbitrary parameters, including those for a conducting medium without any restrictions on vibrator lengths and excitation methods.

The analysis is based on the integrodifferential equation

$$\left(\frac{d^2}{ds^2}+k_1^2\right)\int_{-L}^{L} J(s')\frac{e^{-ik_1R(s,s')}}{R(s,s')}ds' = -i\omega\varepsilon_1 E_{0s}(s) + i\omega\varepsilon_1 z_i J(s), \tag{2.53}$$

where $k_1 = k\sqrt{\varepsilon_1\mu_1} = k_1' - ik_1''$ is the wave number in the medium and $\varepsilon_1 \neq 1$, $\mu_1 \neq 1$, $z_i(s) = const$. We obtain the solution of (2.53) as we did that of (2.1), i.e., by the change of variables

$$\begin{aligned} J(s) &= A(s)\cos k_1 s + B(s)\sin k_1 s,\\ \frac{dJ(s)}{ds} &= -A(s)k_1 \sin k_1 s + B(s)k_1 \cos k_1 s. \end{aligned} \tag{2.54}$$

Using the methods described in Sect. 2.1, we obtain an approximate expression for the current in a thin impedance vibrator located in an infinite homogeneous lossy medium:

$$\begin{aligned} J(s) = \alpha\frac{i\omega\varepsilon_1}{k_1}\Bigg\{&\int_{-L}^{s} E_{0s}(s')\sin\tilde{k}_1(s-s')ds' \\ &- \frac{\sin\tilde{k}_1(L+s) + \alpha P^s[k_1 r, \tilde{k}_1(L+s)]}{\sin 2\tilde{k}_1 L + \alpha P^s(k_1 r, 2\tilde{k}_1 L)}\int_{-L}^{L} E_{0s}^s(s')\sin\tilde{k}_1(L-s')ds' \\ &- \frac{\sin\tilde{k}_1(L+s) + \alpha P^a[k_1 r, \tilde{k}_1(L+s)]}{\sin 2\tilde{k}_1 L + \alpha P^a(k_1 r, 2\tilde{k}_1 L)}\int_{-L}^{L} E_{0s}^a(s')\sin\tilde{k}_1(L-s')ds'\Bigg\}. \end{aligned} \tag{2.55}$$

Here $\tilde{k}_1 = k_1 + i(\alpha/r)\overline{Z}_S\sqrt{\varepsilon_1/\mu_1}$, $G(s,s') = \dfrac{e^{-ik_1\sqrt{(s-s')^2+r^2}}}{\sqrt{(s-s')^2+r^2}}$,

$$\begin{aligned} P^s[k_1 r, \tilde{k}_1(L+s)] &= \int_{-L}^{s}[G(s',-L)+G(s',L)]\sin\tilde{k}_1(s-s')ds'\Big|_{s=L} \\ &= P^s(k_1 r, 2\tilde{k}_1 L), \\ P^a[k_1 r, \tilde{k}_1(L+s)] &= \int_{-L}^{s}[G(s',-L)-G(s',L)]\sin\tilde{k}_1(s-s')ds'\Big|_{s=L} \\ &= P^a(k_1 r, 2\tilde{k}_1 L). \end{aligned} \tag{2.56}$$

If the vibrator is excited by a lumped EMF at the center, the expression for the current (2.55) has the form

$$J(s) = -\alpha V_0\left(\frac{i\omega\varepsilon_1}{2\tilde{k}_1}\right)\frac{\sin\tilde{k}_1(L-|s|) + \alpha P_\delta^s(k_1 r, \tilde{k}_1 s)}{\cos\tilde{k}_1 L + \alpha P_L^s(k_1 r, \tilde{k}_1 L)}. \tag{2.57}$$

Here the equality $P^{s}_{\delta}(k_1 r,\tilde{k}_1 s) = P^{s}[k_1 r,\tilde{k}_1(L+s)] - (\sin\tilde{k}_1 s + \sin\tilde{k}_1|s|)P^{s}_{L}(k_1 r,\tilde{k}_1 L)$, $P^{s}[k_1 r,\tilde{k}_1(L+s)]$ is defined by (2.56), and $P^{s}_{L}(k_1 r,\tilde{k}_1 L) = \int\limits_{-L}^{L} G(s,L)\cos\tilde{k}_1 s\,ds$.

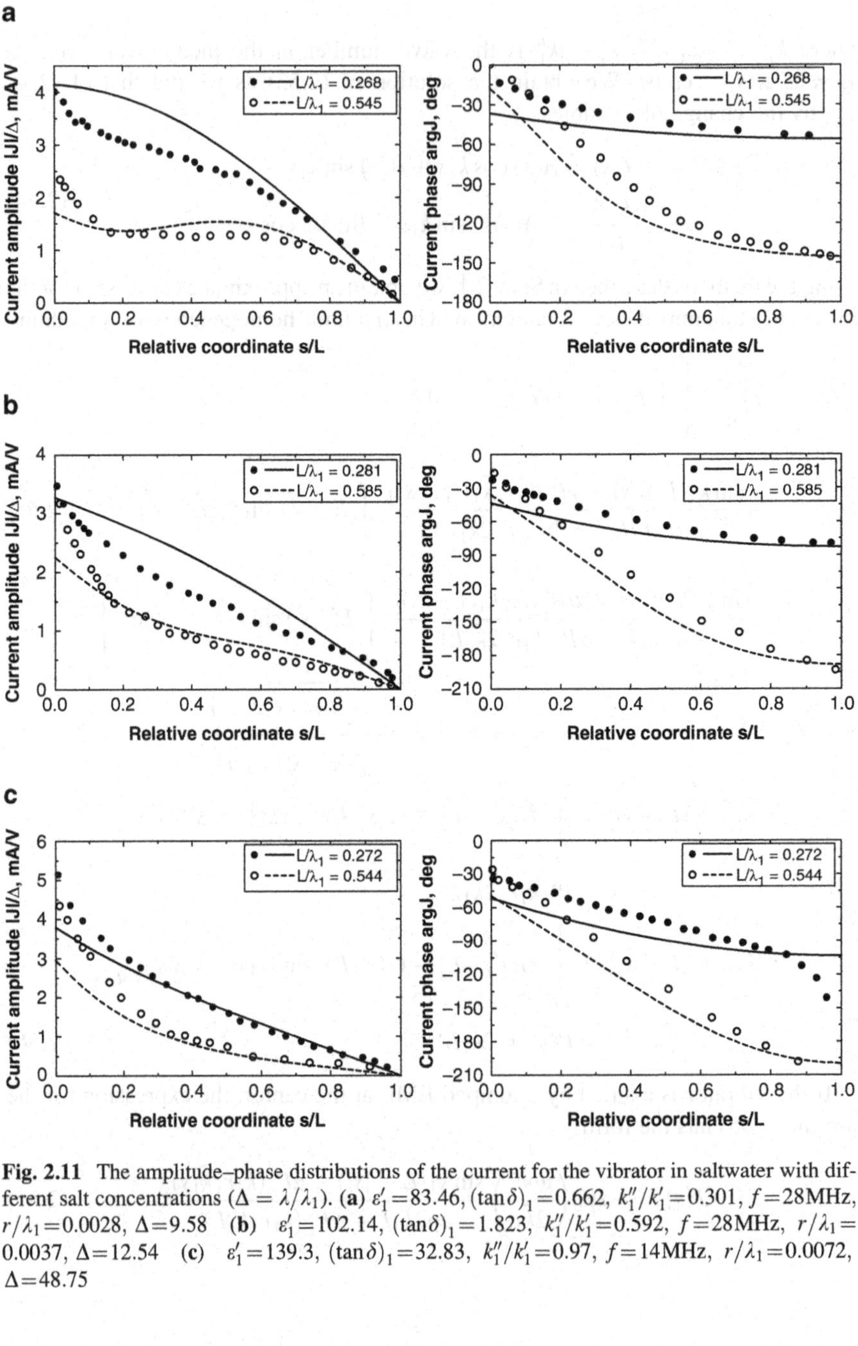

Fig. 2.11 The amplitude–phase distributions of the current for the vibrator in saltwater with different salt concentrations ($\Delta = \lambda/\lambda_1$). **(a)** $\varepsilon_1' = 83.46$, $(\tan\delta)_1 = 0.662$, $k_1''/k_1' = 0.301$, $f = 28\text{MHz}$, $r/\lambda_1 = 0.0028$, $\Delta = 9.58$ **(b)** $\varepsilon_1' = 102.14$, $(\tan\delta)_1 = 1.823$, $k_1''/k_1' = 0.592$, $f = 28\text{MHz}$, $r/\lambda_1 = 0.0037$, $\Delta = 12.54$ **(c)** $\varepsilon_1' = 139.3$, $(\tan\delta)_1 = 32.83$, $k_1''/k_1' = 0.97$, $f = 14\text{MHz}$, $r/\lambda_1 = 0.0072$, $\Delta = 48.75$

To validate approximate analytical solution for the current (2.57), Fig. 2.11 shows the amplitude–phase distributions of the current for the perfectly conducting vibrator ($\overline{Z}_S = 0$), calculated and plotted for different vibrator lengths and medium absorption coefficients, together with the experimental values from [5] (the circles). Since theoretical and experimental curves agree well, we may conclude that the proposed mathematical model corresponds to the real electromagnetic process.

As we may suppose, the electrophysical parameters of the environment influence sufficiently the amplitude–phase distributions of the current in the vibrator. This can be proved by the plots in Fig. 2.12, where are shown curves for the normalized amplitude $|J(s)|/|J|_{\max}$ and the current phase $\arg J(s)$ along the arm of a symmetric perfectly conducting half-wave vibrator located in a biological medium with the electrophysical parameters given in Table 2.1.

Figure 2.12 demonstrates the variation in the electrical length of a vibrator $2L/\lambda_1$ in a material medium, proved by additional extrema and the sections with opposite phases in the distributions of the current along the vibrator, and this variation increases with the density of the medium.

2.4 Radiation Fields of Impedance Vibrators in Infinite Medium

Expressions (2.57), (1.3), and (1.12) define the radiation fields of a thin impedance vibrator in a material medium. These fields may be written in spherical coordinates ρ, θ, φ (θ is the angle measured from the vibrator axis) as

$$
\begin{aligned}
E_\rho(\rho,\theta) &= \frac{k_1}{\omega\varepsilon_1}\int\limits_{-L}^{L} J(s)\frac{e^{-ik_1R(s)}}{R^3(s)}\left\{\begin{matrix} 2R(s)\left[1+\frac{1}{ik_1R(s)}\right]\cos\theta \\ -ik_1\rho\left[1+\frac{3}{ik_1R(s)}-\frac{3}{k_1^2R^2(s)}\right]s\sin^2\theta \end{matrix}\right\}ds,\\
E_\theta(\rho,\theta) &= -\frac{k_1\sin\theta}{\omega\varepsilon_1}\int\limits_{-L}^{L} J(s)\frac{e^{-ik_1R(s)}}{R^3(s)}\left\{\begin{matrix} 2R(s)\left[1+\frac{1}{ik_1R(s)}\right] \\ -ik_1\rho\left[1+\frac{3}{ik_1R(s)}-\frac{3}{k_1^2R^2(s)}\right](\rho - s\cos\theta) \end{matrix}\right\}ds,\\
H_\varphi(\rho,\theta) &= \frac{ik_1k\sin\theta}{\omega}\int\limits_{-L}^{L} J(s)\frac{e^{-ik_1R(s)}}{R^2(s)}\left[1+\frac{1}{ik_1R(s)}\right]\rho\, ds,\\
E_\varphi(\rho,\theta) &= H_\rho(\rho,\theta) = H_\theta(\rho,\theta) = 0, \quad R(s) = \sqrt{\rho^2 - 2\rho s\cos\theta + s^2},
\end{aligned}
\tag{2.58}
$$

and the power absorbed in a unit volume of dielectric is given by

$$
\bar{P}(\rho,\theta) \sim |\vec{E}(\rho,\theta)|^2\omega\varepsilon_1'', \tag{2.59}
$$

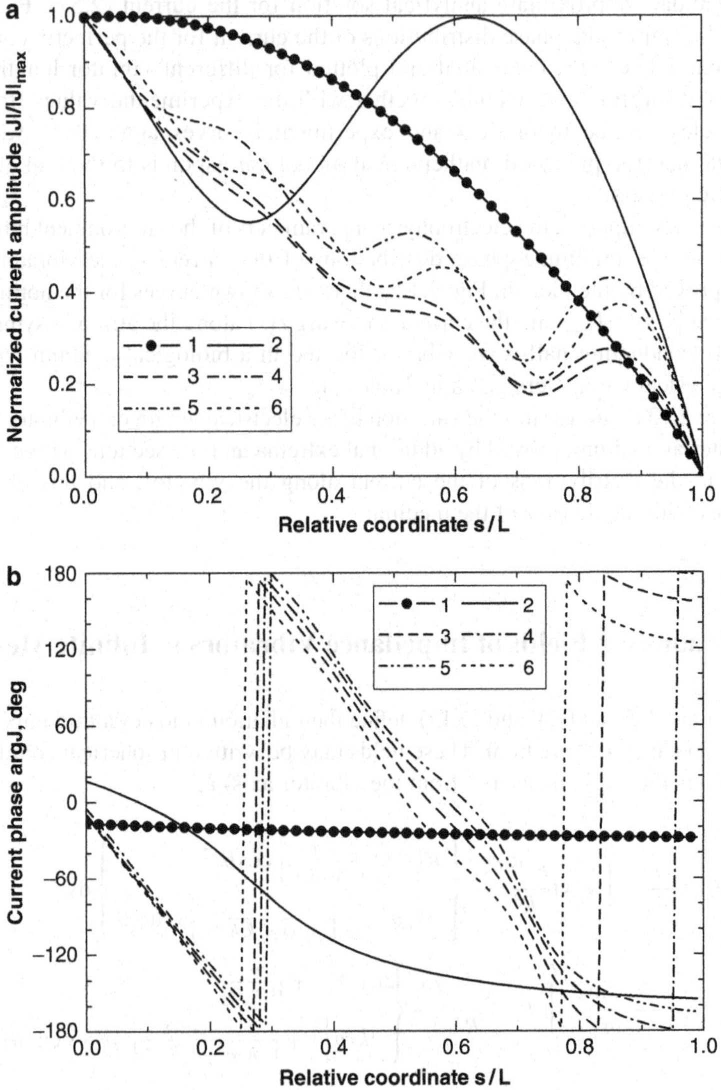

Fig. 2.12 The amplitude–phase distributions of the current for a perfectly conducting vibrator in different medium for $r/\lambda = 0.007022$ and $2L/\lambda = 0.5$: *1* free space; *2* fat layer; *3* muscular tissue; *4* skin; *5* liver; *6* whole blood

Table 2.1 The electrophysical parameters ($\varepsilon_1 = \varepsilon_1' - i\varepsilon_1''$, $(\tan\delta)_1 = \varepsilon_1''/\varepsilon_1'$) for human body tissues (the wavelength is $\lambda = 10$ cm and the temperature is 37°C [12])

Medium	ε_1'	ε_1''	$(\tan\delta)_1$
Free space	1.0	0.0	0.0
Fat layer	6.5	1.6	0.246
Muscular tissue	46.5	18.0	0.387
Skin	43.5	16.5	0.379
Liver	42.5	12.2	0.287
Whole blood	53.0	15.0	0.283

where $\vec{E}(\rho,\theta)=\vec{e}_\rho E_\rho(\rho,\theta)+\vec{e}_\theta E_\theta(\rho,\theta)$, $\varepsilon_1''=4\pi\sigma_1/\omega$, σ_1 is the medium conductivity, and $\vec{e}_\rho$ and $\vec{e}_\theta$ are unit vectors.

Expressions for the fields of an electrically short vibrator (dipole) in a homogeneous isotropic lossy medium for $|k_1L|\ll 1$ may be derived from (2.58) with $J(s)=J_0$ and $R(s)\approx\rho$:

$$E_\rho(\rho,\theta)=-\mathrm{i}2LJ_0\frac{2k_1^2\cos\theta\,\mathrm{e}^{-\mathrm{i}k_1\rho}}{\omega\varepsilon_1\rho}\left(\frac{1}{k_1^2\rho^2}+\frac{\mathrm{i}}{k_1\rho}\right), \tag{2.60a}$$

$$E_\theta(\rho,\theta)=-\mathrm{i}2LJ_0\frac{k_1^2\sin\theta\,\mathrm{e}^{-\mathrm{i}k_1\rho}}{\omega\varepsilon_1\rho}\left(\frac{1}{k_1^2\rho^2}+\frac{\mathrm{i}}{k_1\rho}-1\right), \tag{2.60b}$$

$$H_\varphi(\rho,\theta)=-\mathrm{i}2LJ_0\frac{k_1k\sin\theta\,\mathrm{e}^{-\mathrm{i}k_1\rho}}{\omega\rho}\left(\frac{\mathrm{i}}{k_1\rho}-1\right). \tag{2.60c}$$

The structure of the electromagnetic field in the immediate vicinity of the vibrator is rather complex. However, for $\rho\to\infty$ and $\rho\gg 2L$ ($R(s)\cong\rho-s\cos\theta$) we may substitute in (2.58)

$$\frac{1}{R(s)}\cong\frac{1}{\rho},\qquad \mathrm{e}^{-\mathrm{i}k_1R(s)}\cong\mathrm{e}^{-\mathrm{i}k_1\rho}\,\mathrm{e}^{\mathrm{i}k_1s\cos\theta}, \tag{2.61}$$

and for $|k_1\rho|\to\infty$, the radiation field has the form

$$\begin{aligned}E_\theta(\rho,\theta)&=\frac{\mathrm{i}k_1^2}{\omega\varepsilon_1}\sin\theta\frac{\mathrm{e}^{-\mathrm{i}k_1\rho}}{\rho}\int\limits_{-L}^{L}J(s)\mathrm{e}^{\mathrm{i}k_1s\cos\theta}\,\mathrm{d}s,\\ H_\varphi(\rho,\theta)&=\frac{\mathrm{i}k_1k}{\omega}\sin\theta\frac{\mathrm{e}^{-\mathrm{i}k_1\rho}}{\rho}\int\limits_{-L}^{L}J(s)\mathrm{e}^{\mathrm{i}k_1s\cos\theta}\,\mathrm{d}s,\end{aligned} \tag{2.62}$$

and the characteristic impedance of the medium becomes $E_\theta/H_\varphi=\sqrt{\mu_1/\varepsilon_1}$.

In Fig. 2.13 (here and below, $\lambda=10$ cm, $r/\lambda=0.0033$) are shown the normalized amplitudes $|\overline{E}_s|^2=|E_s|^2/|E_s|^2_{\max}$ for the field parallel to the axis of a half-wave vibrator, $E_s(\rho,\theta)=E_\rho(\rho,\theta)\cos\theta-E_\theta(\rho,\theta)\sin\theta$, as a function of surface impedance for different environmental parameters. As may be seen, the resonant tuning ($\tilde{k}_1L\cong\pi/2$) requires that the distributed impedance of the vibrator in the lossy material should transit from capacitive ($\overline{X}_S<0$) to inductive ($\overline{X}_S>0$) type and that resonance should occur for higher values $\overline{X}_S$ when ε_1' and ε_1'' are increased.

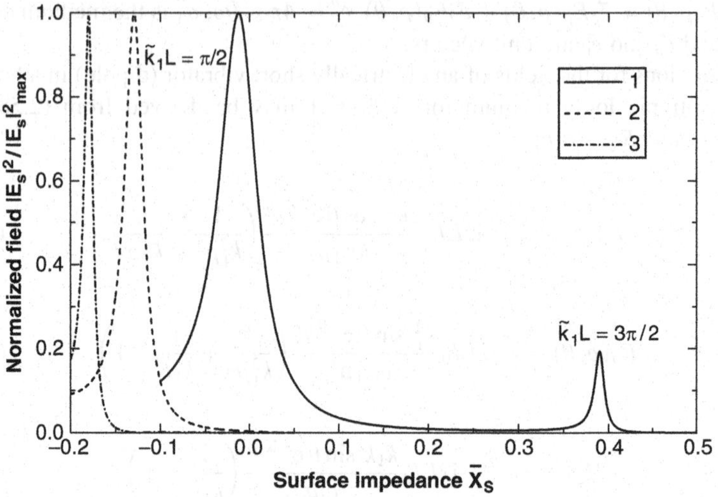

Fig. 2.13 Near-field amplitude versus the surface impedance of the vibrator for $kL = \pi/2$, $\rho/\lambda = 0.5, \theta = 90°$: *1* free space; *2* fat layer; *3* muscular tissue

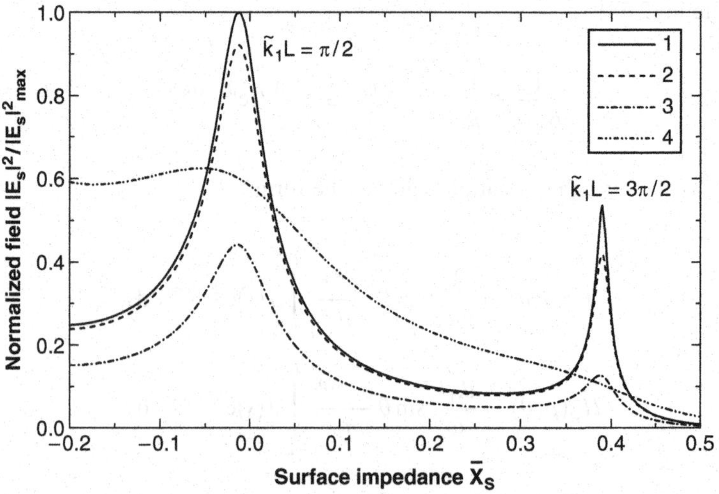

Fig. 2.14 The field amplitude in the far zone versus the surface impedance of the vibrator for $kL = \pi/2, \rho/\lambda = 10.0, \theta = 90°$: *1* $\overline{R}_S = 0.0$; *2* $\overline{R}_S = 0.002$; *3* $\overline{R}_S = 0.01$; *4* $\overline{R}_S = 0.1$

Figure 2.14 represents the dependencies of the far-zone radiation field of a half-wave vibrator in free space upon $\overline{X}_S$ for different real impedances $\overline{R}_S$. As expected, the field decreases in comparison to that of a perfectly conducting vibrator when $\overline{R}_S$ is increased. Note that the vibrator could not be tuned for large values of $\overline{R}_S (\overline{R}_S \geqslant 0.1)$.

The electrophysical parameters of the environment influence considerably the spatial distribution of the electromagnetic field radiated by the vibrator, and thus the absorbed power in the unit volume. This conclusion can be reached by analyzing the plots in Figs. 2.15–2.17, where the distributions of the radiation field for a half-wave impedance vibrator at different distances from its axis are presented for the resonant $\tilde{k}_1 L$ values in free space, the fat layer, and human muscular tissue at 37°C.

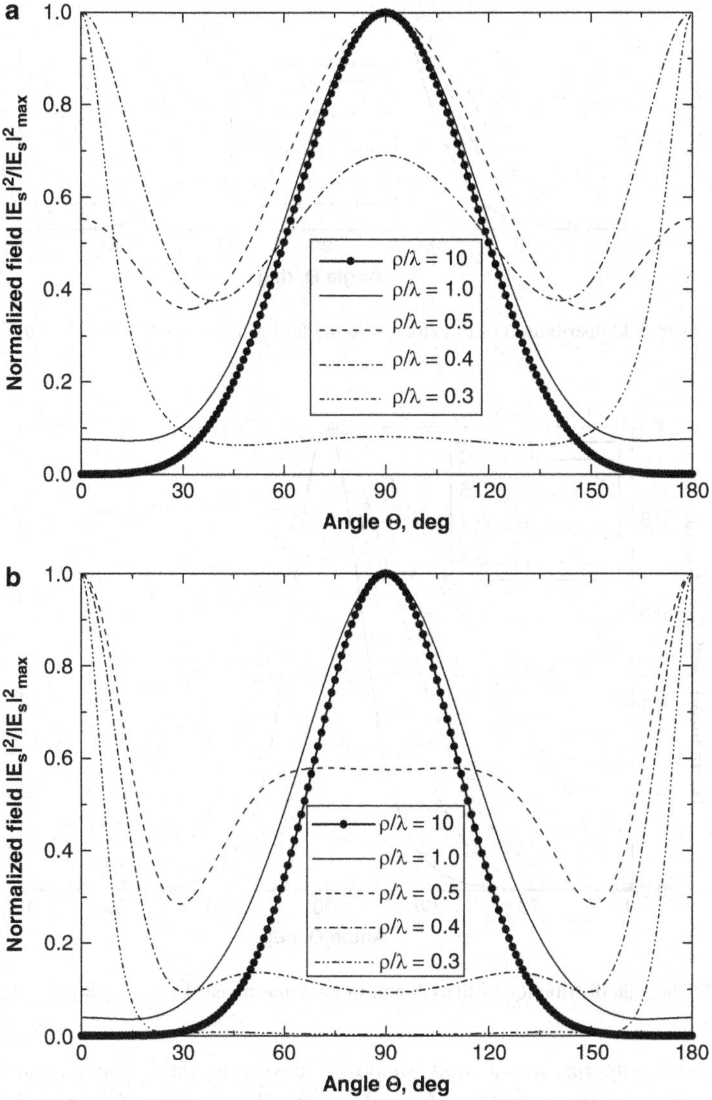

Fig. 2.15 The distribution of the radiation field of a vibrator in free space: (**a**) $\overline{X}_S = -0.013(\tilde{k}_1 L = 0.47\pi)$, (**b**) $\overline{X}_S = 0.39(\tilde{k}_1 L = 1.44\pi)$

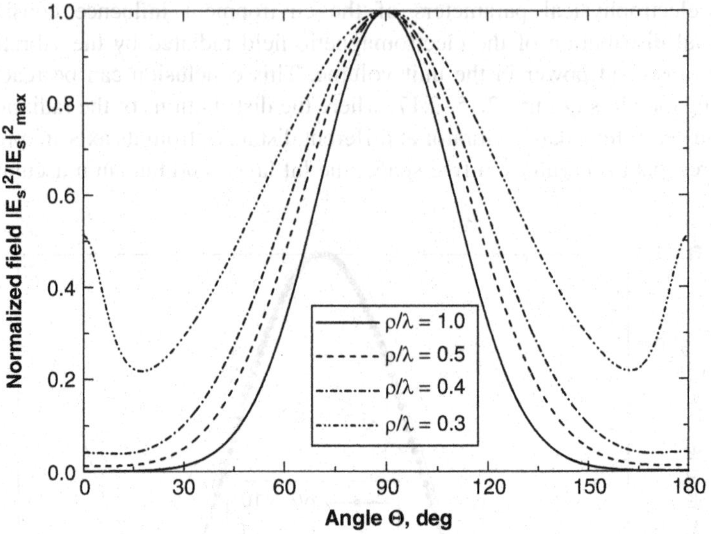

Fig. 2.16 The field distribution of the vibrator in the fat layer: $\overline{X}_S = -0.129(\tilde{k}_1 L = 0.47\pi)$

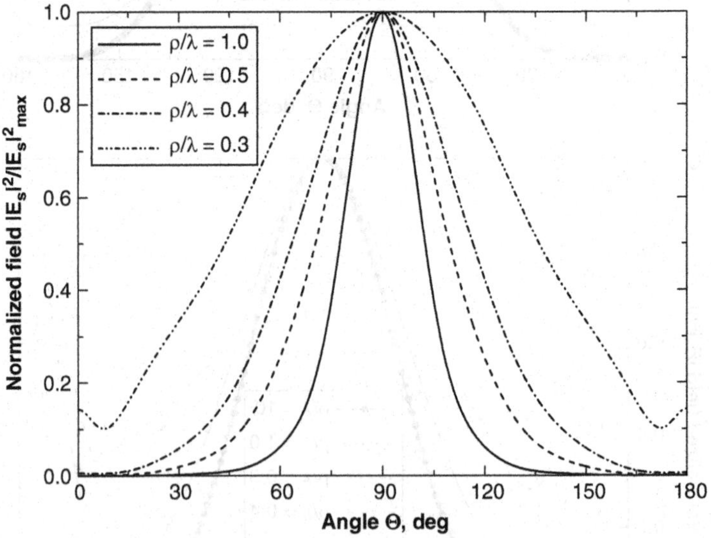

Fig. 2.17 The field distribution of the vibrator in muscular tissue: $\overline{X}_S = -0.18(\tilde{k}_1 L = 0.47\pi)$

The wire antennas are usually made of lossy materials, that is, the surface impedance is really complex, $\overline{Z}_S = \overline{R}_S + i\overline{X}_S$. The values $\overline{Z}_S$ depend on the operating wavelength λ and the vibrator radius r, as shown in Sect. 2.2.2. Figure 2.18 demonstrates the dependence of the normalized field $|\overline{E}|^2$ for the vibrator radius

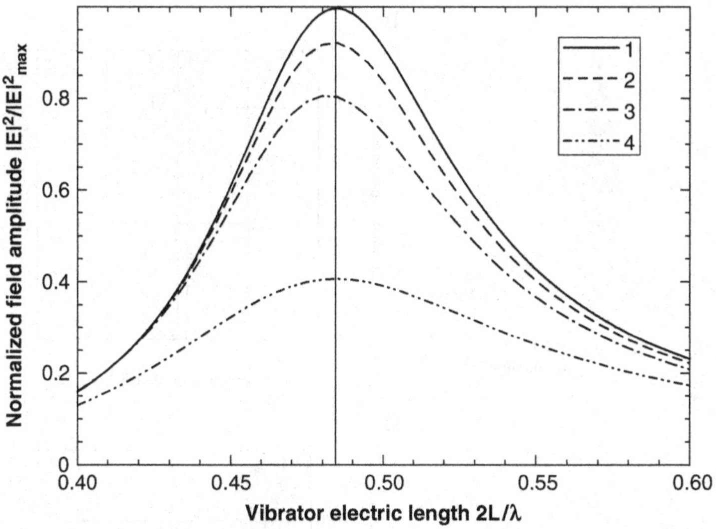

Fig. 2.18 The radiation field of the vibrator $|\overline{E}|^2$ versus electrical length $2L/\lambda$ for different materials for $\rho = \lambda$, $\theta = \pi/2$: *1* $z_i = 0$ [ohm/m] for a perfect conductor; *2* $z_i = 189 + \mathrm{i}180$ [ohm/m] for copper; *3* $z_i = 527 + \mathrm{i}458$ [ohm/m] for platinum; *4* $z_i = 2940 + \mathrm{i}700$ [ohm/m] for bismuth. The impedance values are taken from [13]

$r = 0.00127$ cm at distance $\rho = \lambda$ along the vibrator normal $(\theta = \pi/2)$ at wavelength $\lambda = 10$ cm upon the electrical length of vibrators made of different materials. As is evident from the graph, the vibrator material does not practically influence its resonant length, while the vibrator radiation efficiency decreases substantially as the active component of impedance $\overline{R}_S$ increases.

This is also proved by the surface and contour plots in Fig. 2.19a, b, where the values, normalized by the maximal value of $|\overline{E}|^2$, for a half-wave vibrator versus the active $\overline{R}_S$ and the reactive $\overline{X}_S$ parts of its surface impedance are presented. Let us note that as $\overline{R}_S$ increases, the bandwidth of the value $\overline{X}_S$ where the vibrator is efficiently excited widens, reducing the requirements to the specification of $\overline{R}_S$ and $\overline{X}_S$ for an antenna with artificial complex impedance. In our calculations, the active part of the vibrator impedance was equated to $\overline{R}_S = 0.001$, since this value gives good coincidence between our formulas and real electrodynamic processes.

Since the resonant length of the vibrator essentially depends on the imaginary part of the impedance $\overline{X}_S$ (Fig. 2.19c, d), a quarter-wave ($kL = \pi/4, 2L = \lambda/4$) may be used instead of a half-wave vibrator by choosing $\overline{X}_S$ for the material medium so that $\tilde{k}_1 L = \pi/2$ (Fig. 2.20), thus allowing antenna miniaturization for some applications. The spatial distribution of the normalized near fields for half-wave and quarter-wave vibrators in free space are shown in Fig. 2.21. As can be seen, the radiation field of the quarter-wave vibrator is more homogeneous than that of the half-wave vibrator at small distances. However, it decreases faster when the distance is increased.

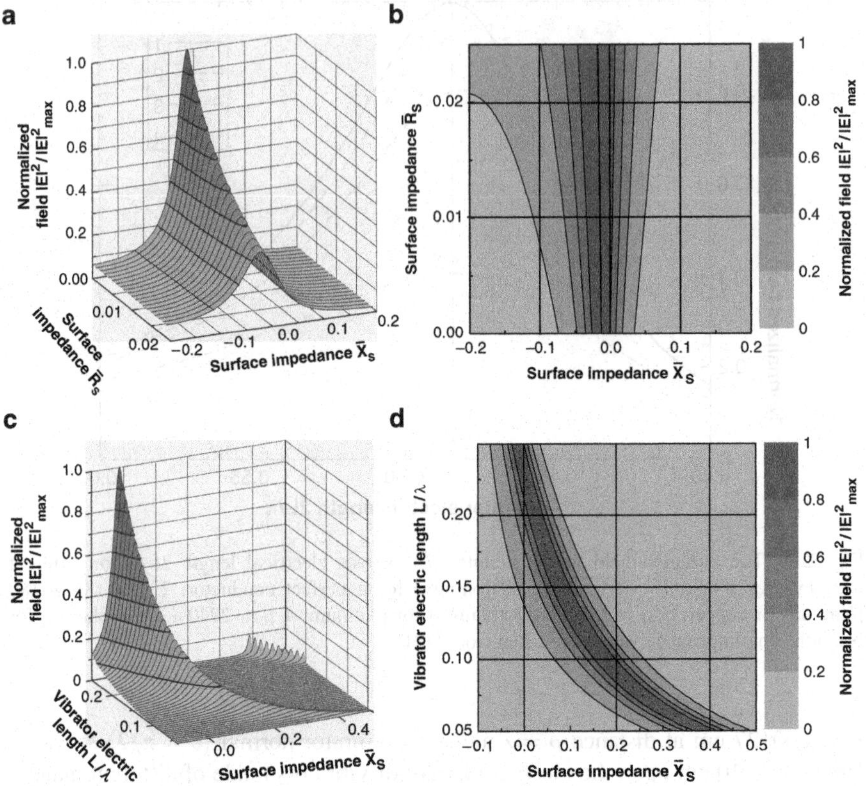

Fig. 2.19 The value of $|\overline{E}|^2$ for a vibrator in unbounded space versus active $\overline{R}_S$ and reactive $\overline{X}_S$ components of the surface impedance

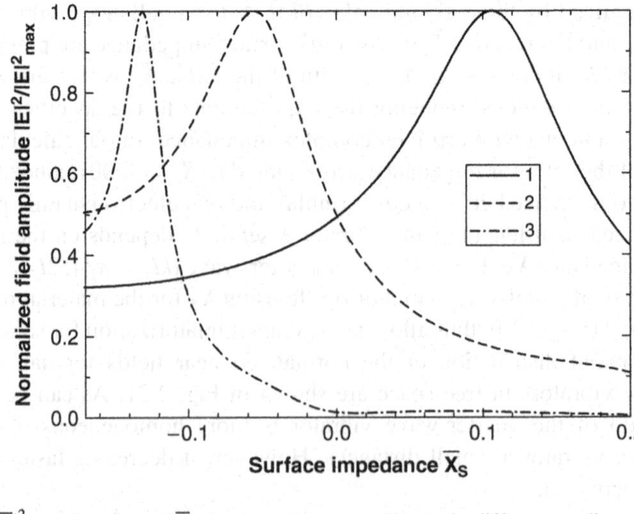

Fig. 2.20 $|\overline{E}|^2$ as a function of $\overline{X}_S$ for a quarter-wave vibrator in different medium: *1* free space; *2* fat layer; *3* muscular tissue

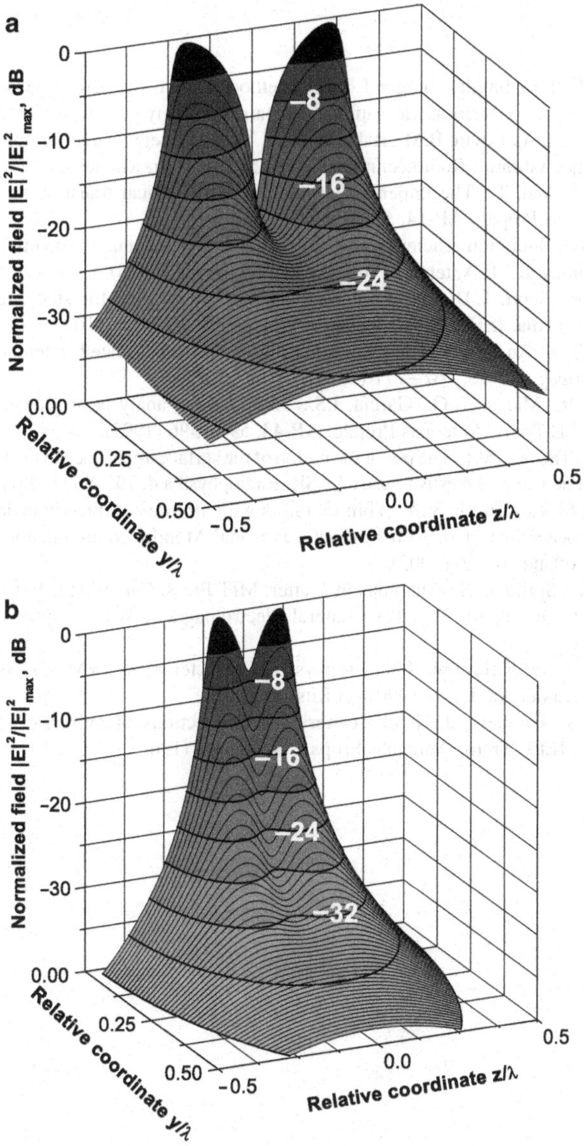

Fig. 2.21 The field distribution of a vibrator in free space: (**a**) $kL = \pi/2$, $\overline{X}_S = -0.013$; (**b**) $kL = \pi/4$, $\overline{X}_S = 0.121$

In conclusion, it may be emphasized that the model of infinite space may be applied to the analysis of impedance vibrators in real medium having finite dimensions, since excited fields in lossy medium rapidly decay as distance from the vibrator increases. A vibrator field near a metallic plane has quite different properties, and its analysis is the subject of the next chapter.

References

1. Kamke, E.: Differentialgleichungen Lösungsmethoden und Lösungen. I. Gewöhnliche Differentialgleichungen. 6. Verbesserte Auflage, Leipzig, Germany (1959) (in German).
2. Glushkovskiy, E.A., Levin, B.M., Rabinovich, E.Y.: The integral equation for the current in a thin impedance vibrator. Radiotechnika **22**, 18–23 (1967) (in Russian).
3. King, R.W.P., Wu, T.: The imperfectly conducting cylindrical transmitting antenna. IEEE Trans. Antennas Propag. **AP-14**, 524–534 (1966).
4. Hanson, G.W.: Radiation efficiency of nano-radius dipole antennas in the microwave and far-infrared regimes. IEEE Antennas Propag. Mag. **50**, 66–77 (2008).
5. King, R.W.P., Scott, L.D. The cylindrical antenna as a probe for studying the electrical properties of media. IEEE Trans. Antennas Propag. **AP-19**, 406–416 (1971).
6. Lamensdorf, D.: An experimental investigation of dielectric-coated antennas. IEEE Trans. Antennas Propag. **AP-15**, 767–771 (1967).
7. Bretones, A.R., Martin, R.G., García, I.S.: Time-domain analysis of magnetic-coated wire antennas. IEEE Trans. Antennas Propag. **AP-43**, 591–596 (1995).
8. Miller, M.A., Talanov, V.I.: The use of the notion of the surface impedance in the theory of surface electromagnetic waves. Izvestiya vusov USSR. Radiophysika **4**, 795–830 (1961) (in Russian).
9. Nesterenko, M.V., Katrich, V.A.: Thin vibrators with arbitrary surface impedance as handset antennas. Proceedings of the 5th European Personal Mobile Communications Conference. Glasgow, Scotland, 16–20 (2003).
10. King, R.W.P., Smith, G.S.: Antennas in Matter. MIT Press, Cambridge, MA (1981).
11. King, R.W.P., Owens, M., Wu, T.T.: Lateral Electromagnetic Waves. Springer-Verlag, New York (1992).
12. Beresovskiy, V.A., Kolotilov, N.N.: Biophysical Characteristics of a Man's Tissue. Reference book. Naukova dumka, Kiev (1990) (in Russian).
13. Cassedy, E.S., Fainberg, J.: Back scattering cross sections of cylindrical wires of finite conductivity. IEEE Trans. Antennas Propag. **AP-8**, 1–7 (1960).

Chapter 3
Radiation of Electromagnetic Waves by Impedance Vibrators in Material Medium over a Perfectly Conducting Plane

The asymptotic solution of the integral equation for the current in a thin impedance vibrator located in an unbounded homogeneous isotropic lossy medium was obtained in Sect. 2.3. The expressions for the radiation fields as a function of geometrical dimensions, surface impedance of vibrators, and the electrophysical parameters of the environment were derived in that chapter. However, in most practical applications the radiating vibrators are located near metal bodies of different configurations. The simplest configuration is the infinite perfectly conducting plane, which may be a good model approximation for a rectangular or circular plate of finite dimensions.

One of the vibrator antennas widely distributed in practical applications is a monopole antenna consisting of a vertical vibrator element over a plane conducting screen on the interface surface of the medium. Such monopole antennas have a production-friendly construction and are convenient in service due to their simple interface with the coaxial feeder line, where the monopole is an extension of the coaxial cable's central conductor. For an asymmetric vertical vibrator over a perfectly conducting plane, the inclusion of mirror-image currents is equivalent to the addition to the vibrator of a symmetric arm. Thus, the current distribution in the monopole coincides with that in the arm of a symmetric vibrator in an infinite medium, and its input impedance is half that of the symmetric vibrator.

In this section, we obtain approximate analytical formulas for the current and expressions for the electromagnetic fields of a thin horizontal impedance vibrator and a system of crossed vibrators over an infinite perfectly conducting plane in a semi-infinite homogeneous lossy medium. The spatial distribution of the near field and the influence of the perfectly conducting screen on it for these radiators in dependence on the parameters of the material medium have been studied by numerical methods.

M.V. Nesterenko et al., *Thin Impedance Vibrators*, Lecture Notes in Electrical Engineering 2064, DOI 10.1007/978-1-4419-7850-9_3,

3.1 Horizontal Impedance Vibrator in a Semi-infinite Material Medium

The task geometry and related notation are shown in Fig. 3.1. Here $\{x, y, z\}$ and $\{\tilde{x}, \tilde{y}, \tilde{z}\}$ are the Cartesian coordinates related respectively to the plane and the cylindrical vibrator. The vibrator length is $2L$ and its radius is r, $\{\rho, \theta, \varphi\}$ are the spherical coordinates, and the angle θ is measured from the axis s, directed along the vibrator axis. The vibrator is located in the medium with material parameters ε_1, μ_1 at distance h from the plane. The general expression for the current is defined by (2.55), and the function $G_s(s, s')$ for the structure, in accordance with (A.2), is

$$G_s(s, s') = \frac{e^{-ik_1\sqrt{(s-s')^2+r^2}}}{\sqrt{(s-s')^2+r^2}} - \frac{e^{-ik_1\sqrt{(s-s')^2+(2h+r)^2}}}{\sqrt{(s-s')^2+(2h+r)^2}}. \tag{3.1}$$

For the central excitation by a hypothetical generator having voltage V_0, the expression for the vibrator current (2.55) may be written as

$$J(s) = -\alpha V_0 \left(\frac{i\omega\varepsilon_1}{2\tilde{k}_1}\right) \frac{\sin \tilde{k}_1(L-|s|) + \alpha P^s_\delta[k_1(r+h), \tilde{k}_1 s]}{\cos \tilde{k}_1 L + \alpha P^s_L[k_1(r+h), \tilde{k}_1 L]}, \tag{3.2}$$

where $P^s_\delta[k_1(r+h),\tilde{k}_1 s] = P^s[k_1(r+h),\tilde{k}_1(L+s)] - (\sin\tilde{k}_1 s + \sin\tilde{k}_1|s|)P^s_L[k_1(r+h),$ $\tilde{k}_1 L]$ and $P^s[k_1(r+h),\tilde{k}_1(L+s)]$ is defined by (2.56), $\tilde{k}_1 = k_1 + i(\alpha/r)\overline{Z}_S\sqrt{\varepsilon_1/\mu_1}$, and $P^s_L[k_1(r+h),\tilde{k}_1 L] = \int\limits_{-L}^{L} G_s(s,L)\cos\tilde{k}s\,ds$.

The full radiation field of the vibrator in spherical coordinates has six components, but we give explicitly only the expressions for the electrical field

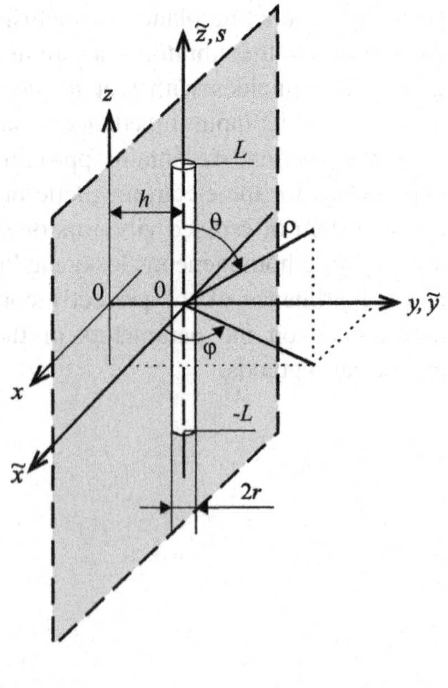

Fig. 3.1 The geometry of the vibrator model

$\vec{E} = \vec{e}_\rho E_\rho + \vec{e}_\theta E_\theta + \vec{e}_\varphi E_\varphi$ ($\vec{e}_\rho, \vec{e}_\theta, \vec{e}_\varphi$ are the unit vectors of the spherical coordinate system):

$$E_\rho(\rho,\theta,\varphi) = \frac{k_1}{\omega\varepsilon_1}\int\limits_{-L}^{L} J(s)\left\{F_\rho[s,R(s)] - F_\rho[s,R_0(s)] - \frac{e^{-ik_1R_0(s)}}{R_0^3(s)}2ik_1h\right.$$
$$\times\left[2h\cos\theta + \sin\varphi\left(s\sin\theta + \frac{1}{2}\rho\sin 2\theta\right)\right]$$
$$\left.\times\left[1+\frac{3}{ik_1R_0(s)} - \frac{3}{k_1^2R_0^2(s)}\right]\right\}ds,$$

$$E_\theta(\rho,\theta,\varphi) = -\frac{k_1}{\omega\varepsilon_1}\int\limits_{-L}^{L} J(s)\left\{F_\theta[s,R(s)] - F_\theta[s,R_0(s)]\right.$$
$$-\frac{e^{-ik_1R_0(s)}}{R_0^3(s)}2ik_1h[2h\sin\theta + \sin\varphi((\rho - s\cos\theta) + \rho\sin^2\theta)]$$
$$\left.\times\left[1+\frac{3}{ik_1R_0(s)} - \frac{3}{k_1^2R_0^2(s)}\right]\right\}ds,$$

$$E_\varphi(\rho,\theta,\varphi) = \frac{k_1}{\omega\varepsilon_1}\int\limits_{-L}^{L} J(s)\frac{e^{-ik_1R_0(s)}}{R_0^3(s)}2ik_1h\cos\varphi(s - \rho\cos\theta)$$
$$\times\left[1+\frac{3}{ik_1R_0(s)} - \frac{3}{k_1^2R_0^2(s)}\right]ds. \tag{3.3}$$

Here $R(s) = \sqrt{\rho^2 - 2\rho s\cos\theta + s^2}$,

$$R_0(s) = \sqrt{\rho^2 - 2\rho s\cos\theta + s^2 + 4h(\rho\sin\theta\sin\varphi + h)},$$

$$F_\rho[s,R_{(0)}(s)] = \frac{e^{-ik_1R_{(0)}(s)}}{R_{(0)}^3(s)}\left\{2R_{(0)}(s)\left[1+\frac{1}{ik_1R_{(0)}(s)}\right]\cos\theta\right.$$
$$\left.-ik_1\rho\left[1+\frac{3}{ik_1R_{(0)}(s)} - \frac{3}{k_1^2R_{(0)}^2(s)}\right]s\sin^2\theta\right\},$$

$$F_\theta[s,R_{(0)}(s)] = \frac{e^{-ik_1R_{(0)}(s)}}{R_{(0)}^3(s)}\sin\theta\left\{2R_{(0)}(s)\left[1+\frac{1}{ik_1R_{(0)}(s)}\right]\right.$$
$$\left.-ik_1\rho\left[1+\frac{3}{ik_1R_{(0)}(s)} - \frac{3}{k_1^2R_{(0)}^2(s)}\right](\rho - s\cos\theta)\right\}.$$

In (3.3), terms with $R(s)$ correspond to the vibrator in infinite space, while terms with $R_0(s)$ represent the perfectly conducting plane. Therefore, the electrical field has all three components, while for an infinite medium, we have $E_\varphi \equiv 0$. Formulas (3.3) are reduced to (2.58) if $h \to \infty$, i.e., the vibrator is in an unbounded medium, and if $\tilde{y} = -h$ $(y = 0)$, the tangential components of the field in the plane become equal to zero; namely, $E_x = (E_\rho \sin\theta + E_\theta \cos\theta)\cos\varphi - E_\varphi \sin\varphi = 0$ and $E_z = E_\rho \cos\theta - E_\theta \sin\theta = 0$.

If the far-zone criteria $\rho \to \infty$, $\rho \gg 2L$, $\rho \gg h$ are satisfied, then $R(s) \approx \rho - s\cos\theta$, $R_0(s) \approx \rho + 2h\sin\theta\,\sin\varphi - s\cos\theta$, $(1/R(s)) = (1/R_0(s)) \approx (1/\rho)$, and $|k_1\rho| \to \infty$, $E_\rho = E_\varphi = 0$, and E_θ is given by

$$E_\theta(\rho,\theta,\varphi) = \frac{\mathrm{i}k_1^2}{\omega\varepsilon_1}\sin\theta\frac{\mathrm{e}^{-\mathrm{i}k_1\rho}}{\rho}(1 - \mathrm{e}^{-2\mathrm{i}k_1 h\sin\theta\sin\varphi})\int_{-L}^{L} J(s)\mathrm{e}^{\mathrm{i}k_1 s\cos\theta}\,\mathrm{d}s. \tag{3.4}$$

The expressions for the field components of an electrically short vibrator (dipole) can be easily derived after substitution in (3.3), $J(s) = J_0$, $R(s) \approx \rho$, and $R_0(s) \approx \rho_h = \sqrt{\rho^2 + 4h(\rho\sin\theta\sin\varphi + h)}$ with $|k_1 L| \ll 1$, as

$$\begin{aligned}
E_\rho(\rho,\theta,\varphi) &= -\mathrm{i}2LJ_0\frac{2k_1^2}{\omega\varepsilon_1}\cos\theta\left\{\frac{\mathrm{e}^{-\mathrm{i}k_1\rho}}{\rho}\left(\frac{1}{k_1^2\rho^2} + \frac{\mathrm{i}}{k_1\rho}\right) - \frac{\mathrm{e}^{-\mathrm{i}k_1\rho_h}}{\rho_h}\left[\left(\frac{1}{k_1^2\rho_h^2} + \frac{\mathrm{i}}{k_1\rho_h}\right)\right.\right.\\
&\quad\left.\left. + \mathrm{i}k_1 h\left(\frac{2h}{\rho} + \sin\theta\,\sin\varphi\right)\times\left(\frac{3\mathrm{i}}{k_1^3\rho_h^3} - \frac{3}{k_1^2\rho_h^2} - \frac{\mathrm{i}}{k_1\rho_h}\right)\right]\right\},\\
E_\theta(\rho,\theta,\varphi) &= -\mathrm{i}2LJ_0\frac{k_1^2}{\omega\varepsilon_1}\left\{\frac{\mathrm{e}^{-\mathrm{i}k_1\rho}}{\rho}\sin\theta\left(\frac{1}{k_1^2\rho^2} + \frac{\mathrm{i}}{k_1\rho} - 1\right)\right.\\
&\quad - \frac{\mathrm{e}^{-\mathrm{i}k_1\rho_h}}{\rho_h}\left[\sin\theta\left(\frac{1}{k_1^2\rho_h^2} + \frac{\mathrm{i}}{k_1\rho_h} - 1\right)\right.\\
&\quad\left.\left. - 2\mathrm{i}k_1 h\left(\frac{2h}{\rho_h}\sin\theta + (1+\sin^2\theta)\sin\varphi\right)\left(\frac{3\mathrm{i}}{k_1^3\rho_h^3} - \frac{3}{k_1^2\rho_h^2} - \frac{\mathrm{i}}{k_1\rho_h}\right)\right]\right\},
\end{aligned} \tag{3.5}$$

$$E_\varphi(\rho,\theta,\varphi) = -\mathrm{i}2LJ_0\frac{k_1^2}{\omega\varepsilon_1}\cos\theta\frac{\mathrm{e}^{-\mathrm{i}k_1\rho_h}}{\rho_h}\left\{2\mathrm{i}k_1 h\cos\varphi\left(\frac{3\mathrm{i}}{k_1^3\rho_h^3} - \frac{3}{k_1^2\rho_h^2} - \frac{\mathrm{i}}{k_1\rho_h}\right)\right\}.$$

The field component satisfies $E_\theta \neq 0$, in contrast to the field of an electrically short vibrator in unbounded space (2.60b) for $\theta = 0^0$; the discrepancy may be explained by the asymmetry of the dipole and its mirror image in the chosen coordinate system.

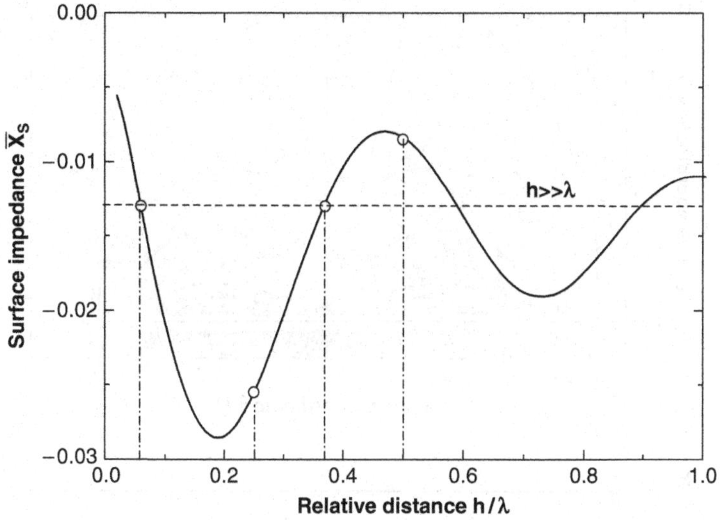

Fig. 3.2 The values of $\overline{X}_{S\,res}$ vs. h/λ

As shown earlier in Sect. 2.2, a vibrator with fixed geometric dimensions, for example the half-wave vibrator ($2L = \lambda/2$) in unbounded space, becomes resonant at a definite value of the imaginary part of its surface impedance $\overline{X}_S$. The values of $\overline{X}_{S\,res}$ as a function of the distance h are represented for a resonant half-wave vibrator over the plane in Fig. 3.2 for $\varepsilon_1 = \mu_1 = 1$. The dotted line corresponds to the case in which the conducting plane is absent ($h \gg \lambda$). As can be seen, the resonant vibrator tuning for different values of h/λ requires matching of the impedance $\overline{X}_{S\,res}$, and there exists a distance h for which $\overline{X}_{S\,res}$ coincides with the value $\overline{X}_{S\,res} = -0.013$, corresponding to a vibrator in unbounded space.

Figures 3.3–3.7 show the spatial distributions of the value $|\vec{E}|$ (in [dB]), normalized by the maximal value at $\tilde{y} = 0.01\lambda$ ($\tilde{x} = 0$) and at $\tilde{y} = 0.1\lambda$, near the half-wave vibrator in two mutually perpendicular planes $\tilde{x} = 0$ and $\tilde{y} = 0.1\lambda$ for $h/\lambda = 0.06$, 0.25, 0.37, 0.5, and ∞. The impedance values required for vibrator resonant tuning and marked by the circles in Fig. 3.2 are given for $\overline{X}_{S\,res} = -0.013$, $-0.026, -0.013, -0.008, -0.013$, respectively. The contour curve for the plane $\tilde{x} = 0$ are plotted with a step size of 5 dB, and for the plane $\tilde{y} = 0.1\lambda$ with a step size of 2 dB. The plots demonstrate that the variation of the distance between a vibrator and a perfectly conducting plane allows one to form various spatial distributions of the near electromagnetic field, while the value of $\overline{X}_{S\,res}$ changes very little. We also note that radiation is more effective for $h = \lambda/4$ (which agrees with (3.4)), and the field is concentrated mainly near the vibrator and in the region between the vibrator and the plane for small values of h/λ and for $h = \lambda/2$.

Thus, we can state that for higher efficiency of a vibrator's radiation, the perfectly conducting plane must be located at $h = \lambda/4$ in free half-space and at $h = n\lambda_1/4$ in material medium. Here λ_1 is the wavelength in the medium, and

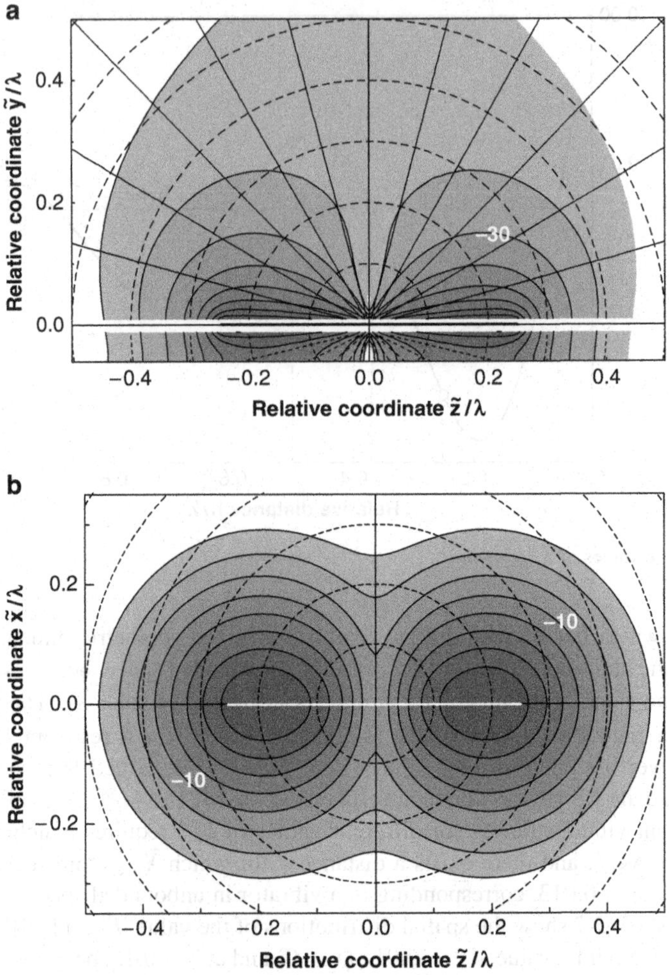

Fig. 3.3 The contour line of $|\vec{E}|$ [dB] for $(h/\lambda) = 0.06$. (**a**) $\tilde{x} = 0$. (**b**) $\tilde{y} = 0.1\lambda$

$n = 1, 3, 5, \ldots$ is the sequence of odd numbers. Thus, in further calculations we used $h = \lambda_1/4$ and $h = 3\lambda_1/4$ for the fat layer and the muscular tissue, respectively. The plots in Fig. 3.8 show that the screen practically does not influence the impedance $\overline{X}_S$, corresponding to the resonant vibrator, and the influence becomes smaller as the medium becomes denser.

The contour plots in Figs. 3.9–3.11 illustrate the influence of the perfectly conducting plane on the distribution of $|\overline{E}|^2$(in [dB]), normalized by the maximal value for each $\tilde{y}/\lambda$ for the quarter-wave $(2L = \lambda/4)$ vibrator in different material medium. In our calculation, the values of impedance $\overline{X}_S$ were selected to ensure the resonant vibrator excitation in the medium according to the maxima in Fig. 3.8.

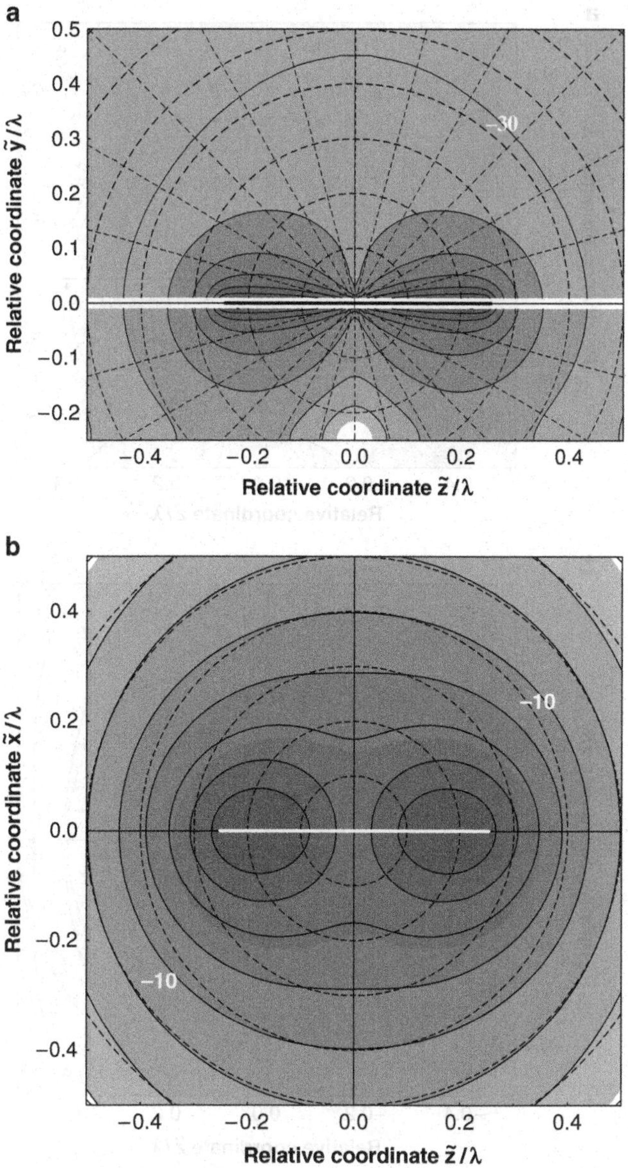

Fig. 3.4 The contour line of $|\vec{E}|$ [dB] for $(h/\lambda) = 0.25$. (**a**) $\tilde{x} = 0$. (**b**) $\tilde{y} = 0.1\lambda$

In Figs. 3.9–3.11a, b, the values for $h \to \infty$ are plotted with solid curves for comparison. The analysis of the plots in Figs. 3.9–3.11 shows that the maximum of the field near a vibrator in a material medium is shifted from the ends to the middle, and this shift grows larger as the medium becomes denser. Let us also

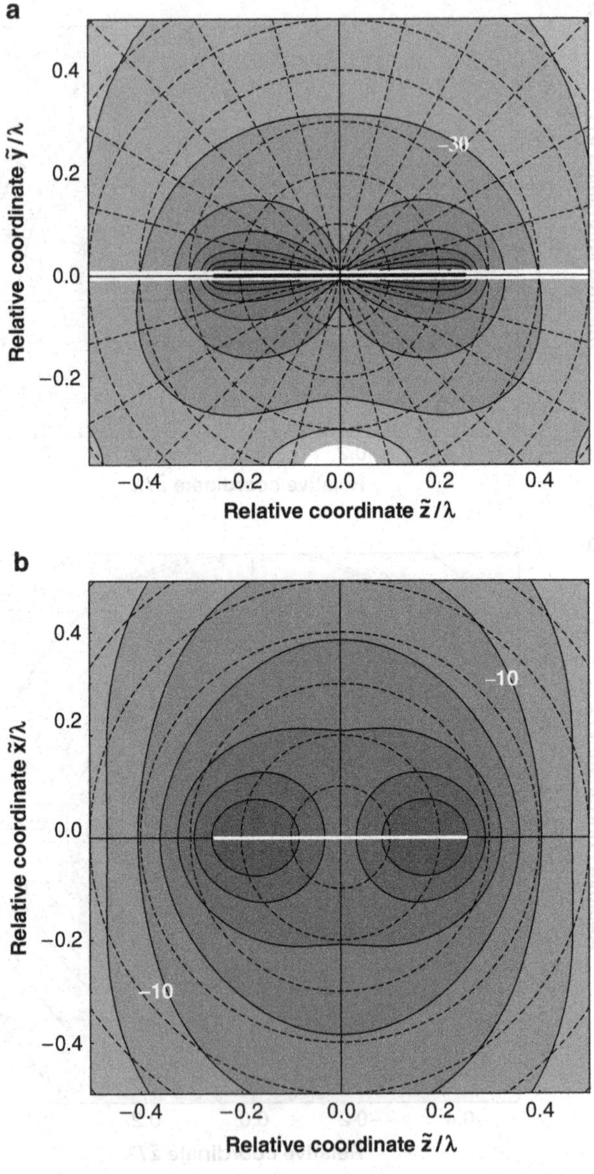

Fig. 3.5 The contour line of $|\vec{E}|$ [dB] for $(h/\lambda) = 0.37$. **(a)** $\tilde{x} = 0$. **(b)** $\tilde{y} = 0.1\lambda$

note that the near field of a vibrator over a plane occupies a larger region than that of a vibrator in unbounded space for $h/\lambda_1 \to \infty$ (see the contour plot in Fig. 3.9). However, boundaries of this region for both vibrator configurations move closer when the medium density is increased, and the spatial concentration of the field increases considerably (Figs. 3.10 and 3.11).

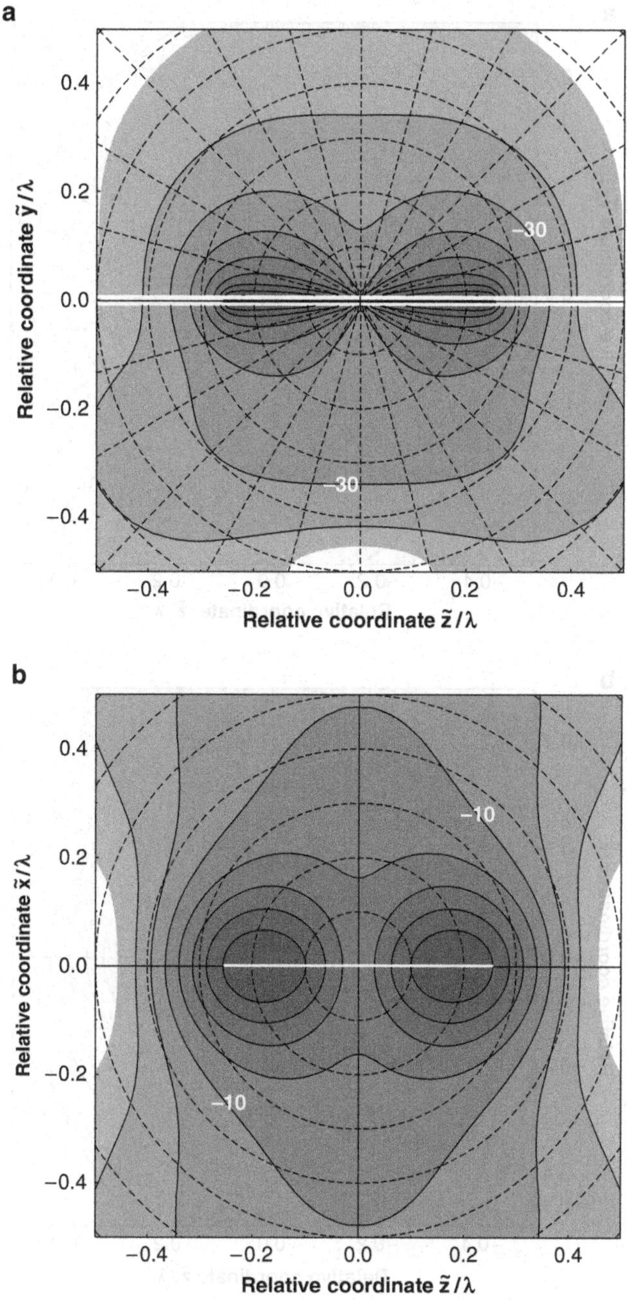

Fig. 3.6 The contour line of $|\vec{E}|$ [dB] for $(h/\lambda) = 0.5$. **(a)** $\tilde{x} = 0$. **(b)** $\tilde{y} = 0.1\lambda$

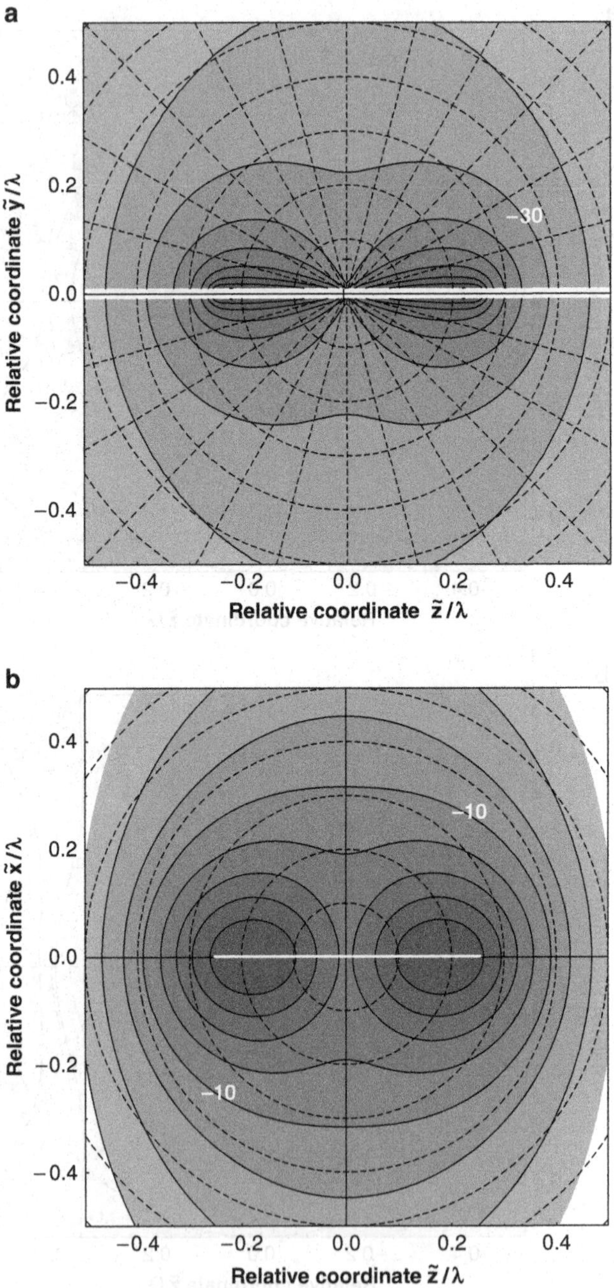

Fig. 3.7 The contour line of $|\vec{E}|$ [dB] for $h \gg \lambda$. (**a**) $\tilde{x} = 0$. (**b**) $\tilde{y} = 0.1\lambda$

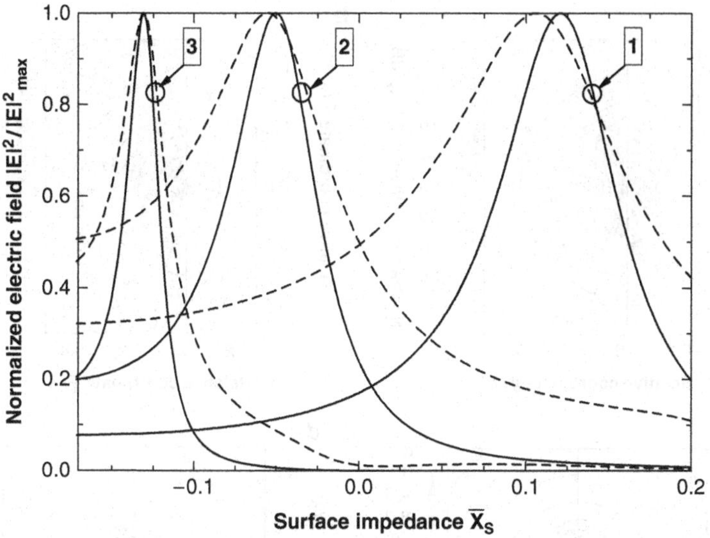

Fig. 3.8 The value of $|\overline{E}|^2$ vs. $\overline{X}_S$ for a quarter-wave vibrator $(2L = \lambda/4)$ in different medium *1* half-space ($h = 2.5$ cm); *2* fat layer ($h = 1.0$ cm); *3* muscular tissue ($h = 1.1$ cm) (the solid lines correspond to $h \to \infty$)

3.2 Systems of Crossed Impedance Vibrators in a Semi-infinite Material Medium

One possible application of microwave radiation in modern clinical medicine is microwave hyperthermia, applied for the controlled deep heating of different tissues and organs of a living organism both near the surface and deep inside it ([10] in Chap. 2, [1–3]). Microwave energy can be delivered to the region to be heated by external radiators (applicators) for subsurface heating or by antenna probes inserted into the organism. The main requirements on the applicators and probes for hyperthermia may be formulated as follows [2]: deep heating with the possibility of focusing and scanning the radiated field, guaranteeing uniform temperature field distribution in heated tissues; good matching with the feeder; small dimensions and weight of applicators and probes. Various antennas, such as a waveguide and horn [4–7], microstrip [8–10], microstrip in combination with slot [11], dielectric waveguide [12], or on circular slots in the outer conductor of the coaxial line, encased in a dielectric shell [13–15], linear [2, 16], and V-shaped [17] vibrators are applied in practice. All microwave antennas and the antennas arrays used for hyperthermia have one common property: they radiate linear polarized waves. To our minds, the efficiency of microwave hyperthermia may be enhanced by application of circular (elliptical) polarized fields, radiated by a system of crossed vibrators with different surface impedances.

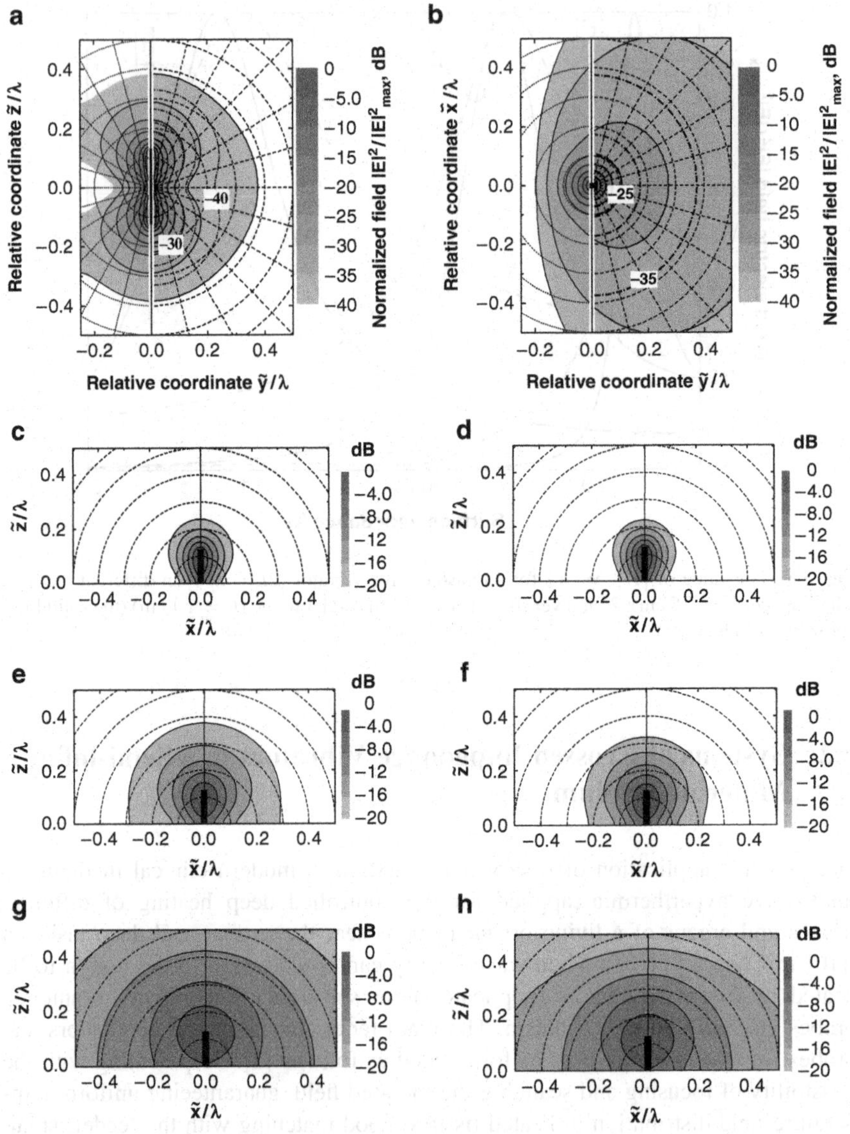

Fig. 3.9 The field distribution for a quarter-wave vibrator in free half-space ($\varepsilon_1 = 1.0$, $\bar{X}_S = 0.121$). (**a**) $h/\lambda_1 = 0.25$, $\tilde{x} = 0$. (**b**) $h/\lambda_1 = 0.25$, $\tilde{z} = 0$. (**c**) $h/\lambda_1 = 0.25$, $\tilde{y}/\lambda = 0.03$. (**d**) $h/\lambda_1 \to \infty$, $\tilde{y}/\lambda = 0.03$. (**e**) $h/\lambda_1 = 0.25$, $\tilde{y}/\lambda = 0.06$. (**f**) $h/\lambda_1 \to \infty$, $\tilde{y}/\lambda = 0.06$. (**g**) $h/\lambda_1 = 0.25, \tilde{y}/\lambda = 0.12$. (**h**) $h/\lambda_1 \to \infty, \tilde{y}/\lambda = 0.12$

The energy and polarization characteristics of a system of two crossed impedance vibrators located in the absorbing medium over a perfectly conducting plane and spatial distributions of the electromagnetic radiation for this system in all observation zones are studied in this section.

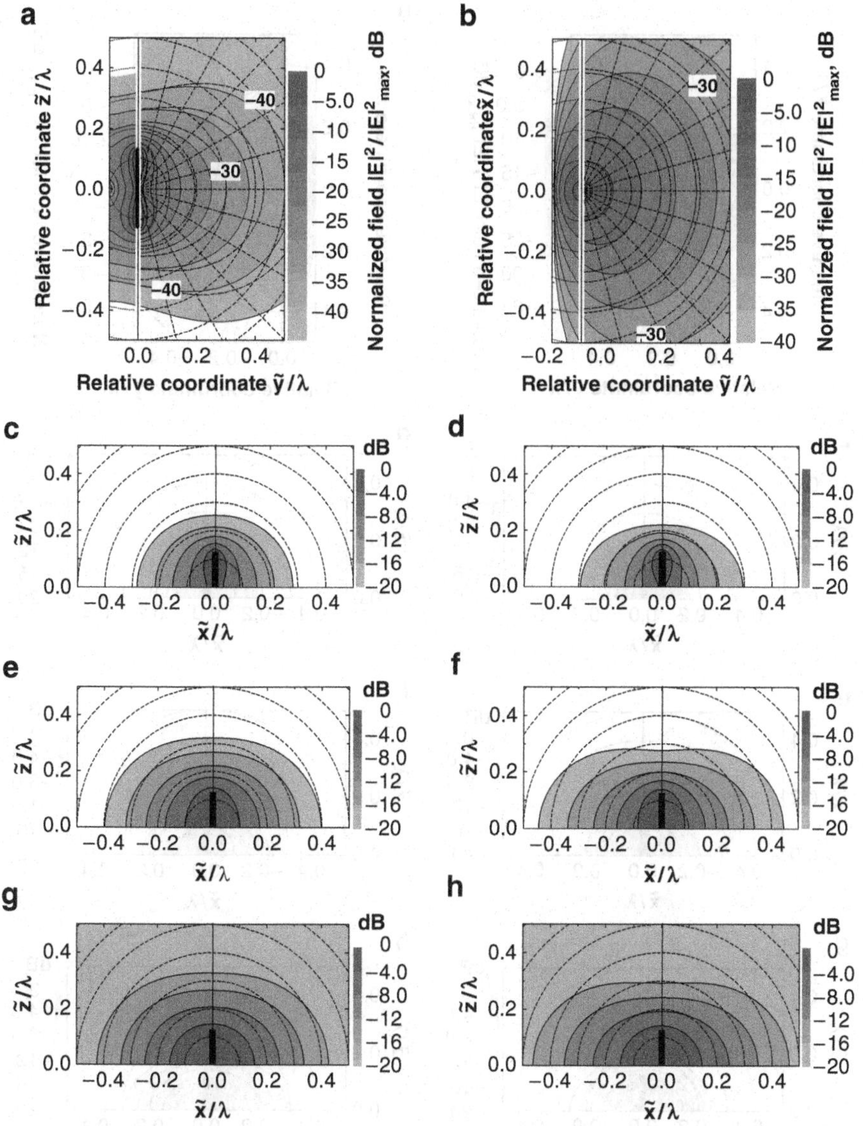

Fig. 3.10 The field distribution for a quarter-wave vibrator in the fat layer ($\varepsilon_1 = 6.5 - i1.6$, $\bar{X}_S = -0.05$). (**a**) $h/\lambda_1 = 0.25$, $\tilde{x} = 0$. (**b**) $h/\lambda_1 = 0.25$, $\tilde{z} = 0$. (**c**) $h/\lambda_1 = 0.25$, $\tilde{y}/\lambda = 0.03$. (**d**) $h/\lambda_1 \to \infty$, $\tilde{y}/\lambda = 0.03$. (**e**) $h/\lambda_1 = 0.25$, $\tilde{y}/\lambda = 0.06$. (**f**) $h/\lambda_1 \to \infty$, $\tilde{y}/\lambda = 0.06$. (**g**) $h/\lambda_1 = 0.25$, $\tilde{y}/\lambda = 0.12$. (**h**) $h/\lambda_1 \to \infty$, $\tilde{y}/\lambda = 0.12$

The geometry of the structure and related notation are shown in Fig. 3.12. Here $\{X, Y, Z\}$ is the Cartesian coordinate system related to crossed cylindrical vibrators with axes $\{0s_n\}$ ($n = 1, 2$) parallel to an infinite perfectly conducting

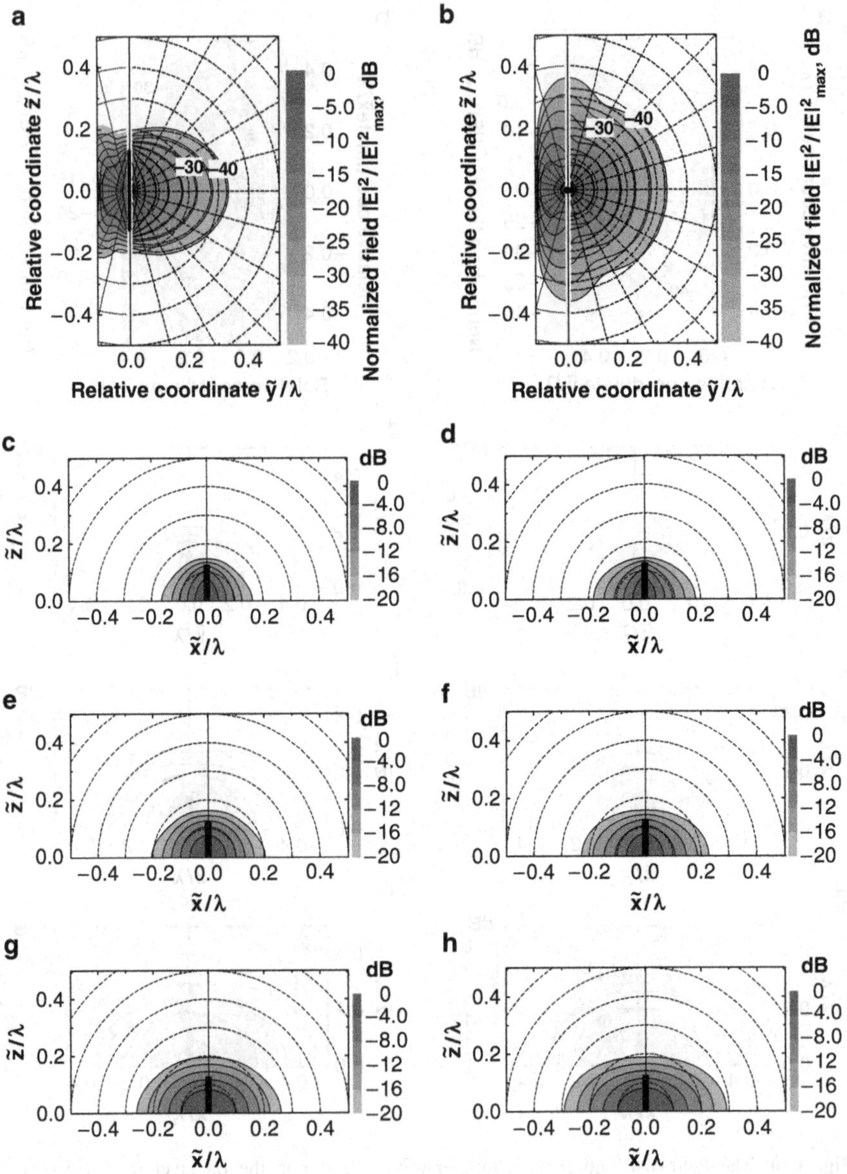

Fig. 3.11 The field distribution for a quarter-wave vibrator in muscular tissue ($\varepsilon_1 = 46.5 - i18.0$, $\bar{X}_S = -0.131$). **(a)** $h/\lambda_1 = 0.75$, $\tilde{x} = 0$. **(b)** $h/\lambda_1 = 0.75$, $\tilde{z} = 0$. **(c)** $h/\lambda_1 = 0.75$, $\tilde{y}/\lambda = 0.03$. **(d)** $h/\lambda_1 \to \infty$, $\tilde{y}/\lambda = 0.03$. **(e)** $h/\lambda_1 = 0.75$, $\tilde{y}/\lambda = 0.06$. **(f)** $h/\lambda_1 \to \infty$, $\tilde{y}/\lambda = 0.06$. **(g)** $h/\lambda_1 = 0.75$, $\tilde{y}/\lambda = 0.12$. **(h)** $h/\lambda_1 \to \infty$, $\tilde{y}/\lambda = 0.12$

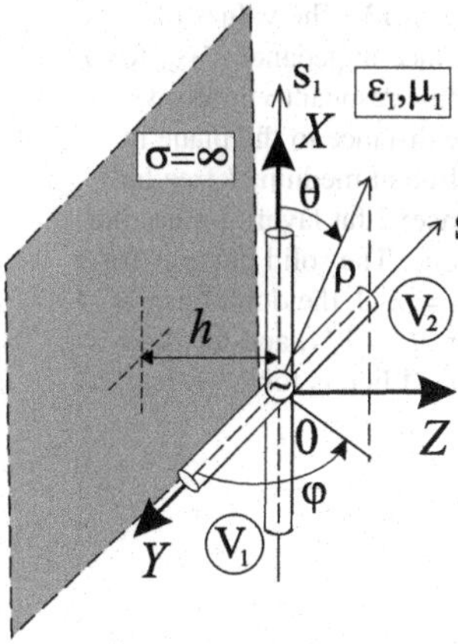

Fig. 3.12 The problem geometry

plane at distance h from it. The vibrators with lengths $2L_n$ and radii r_n $(n = 1, 2)$ are in-phase excited at their centers $(s_n = 0)$ by a hypothetical generator with voltage V_0, and they radiate into the half-space with complex permittivity ε_1 and permeability μ_1.

The electromagnetic field radiated by the system in the spherical coordinate system $\{\rho, \theta, \varphi\}$ has six components, but we shall consider only the electric field components $\vec{E}(\rho,\theta,\varphi) = \vec{E}^{V_1} + \vec{E}^{V_2} = \vec{e}_\rho E_\rho^\Sigma + \vec{e}_\theta E_\theta^\Sigma + \vec{e}_\varphi E_\varphi^\Sigma$, where $\Sigma = \mathrm{V}_1 + \mathrm{V}_2$. The expressions for the fields $E_\rho^{V_1}, E_\theta^{V_1}, E_\varphi^{V_1}$ of a single impedance vibrator located over an infinite screen were obtained in Sect. 3.1, and the vibrator current is given by (3.2). The formula for the current in the second vibrator V_2 is also given by (3.2), and the electrical field components can be determined by (3.3) by coordinate system transformation. The independent solutions of the integral equations for the current in each vibrator can be used here, since in the terms of a "thin-wire" model, the vibrators have different polarizations and do not interact with each other.

This approach may be also used for the analysis of a system consisting of two crossed impedance vibrators in an infinite material medium. The currents in the radiating vibrators may be found by (2.57), and the radiating fields are defined by (2.58). The system of two mutually orthogonal symmetrical perfectly conducting vibrators is known as a turnstile antenna. If the vibrators are fed with equal amplitudes and the phase shift is $\Delta\Psi = \pm(\pi/2)$, the radiation field is circularly polarized in the direction $\theta = \varphi = \pi/2$ perpendicular to the axes of vibrators and has linear polarization in the plane of the vibrators. The radiation field of the turnstile antenna has elliptical polarization at an arbitrary point.

Recall that the resonant tuning of impedance vibrators with fixed dimensions for different medium parameters ε_1 and μ_1, in contrast to perfectly conducting vibrators, can be realized by variation of the imaginary part of the surface impedance. Therefore, the physical dimensions of the crossed vibrators (the lengths of their arms) can be changed arbitrarily if their surface impedance is correctly selected. Figure 3.13 shows the values of the surface impedance $\overline{X}_{S\,\mathrm{res}}$ corresponding to single resonant vibrator vs. the distance h to the plane (the dash-dotted curves

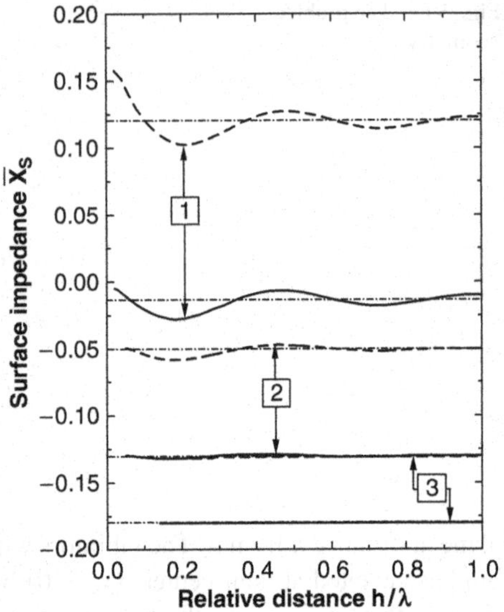

Fig. 3.13 The values of surface impedance $\overline{X}_{S\,res}$ for a single resonant vibrator vs. the distance to the plane in different medium *1* free half-space; *2* fat layer; *3* muscular tissue. The solid curve is for $2L = \lambda/2$, the dotted line is for $2L = \lambda/4$, and the dash-dotted line is for $h \to \infty$

correspond to $h \to \infty$). The dependencies for both the half-wave ($2L = \lambda/2$) and the quarter-wave ($2L = \lambda/4$) vibrators are given here. The plots for both vibrator lengths show that an increase in the medium density reduces the influence of the distance h on the resonant values of the impedance $\overline{X}_{S\,res}$, which for various electric vibrator lengths and medium parameters may be either inductive ($\overline{X}_{S\,res} > 0$) or capacitive ($\overline{X}_{S\,res} < 0$).

The polarization of the radiation field for a system of crossed vibrators in free space can be characterized by the ellipticity r_p and the tilt angle β_p:

$$r_p = \tan\left\{\tfrac{1}{2}\arcsin\frac{2|p|\sin\psi}{|p|^2+1}\right\}, \quad |r_p| \le 1, \quad 0 \le \beta_p < \pi,$$
$$\beta_p = \tfrac{1}{2}\arctan\frac{2|p|\cos\psi}{|p|^2-1}, \tag{3.6}$$

where $|p|$ and ψ are the module and the phase of the ratio $p = (E_\theta^\Sigma(\rho))/(E_\varphi^\Sigma(\rho))$ in the far zone. It is obvious that these expressions can be used for the analysis of the polarization in material medium in the regions where the radial component E_ρ is small and may be neglected. Expressions (3.6) can be used for approximate estimation of the polarization state at other points in space, bearing in mind that the analysis is restricted by the two-component representation.

As indicated earlier, to produce a circular polarized radiation field with a classical turnstile antenna, the constant phase shift $\Delta\Psi = \pm(\pi/2)$ must be incorporated into the excitation scheme for either of two vibrators. The elliptically (circularly) polarized radiation may also be formed with other $\Delta\Psi$ values by a simultaneous change of

Fig. 3.14 The ellipticity of the radiation field for crossed vibrators vs. surface impedances (for free half-space in the far zone with $2L = \lambda/2$, $h = \lambda/4$): (**a**) $r_p(\overline{X}_{S1}, \overline{X}_{S2})$, (**b**) $r_p(\overline{X}_{S2})$: 1–$\overline{X}_{S1} = -0.026$, $\Delta\Phi = 0$; 2–$\overline{X}_{S1} = -0.026$, $\Delta\Phi = \pi/2$; 3–$\overline{X}_{S1} = -0.0606$, $\Delta\Phi = 0$; 4–$\overline{X}_{S1} = 0.0092$, $\Delta\Phi = 0$

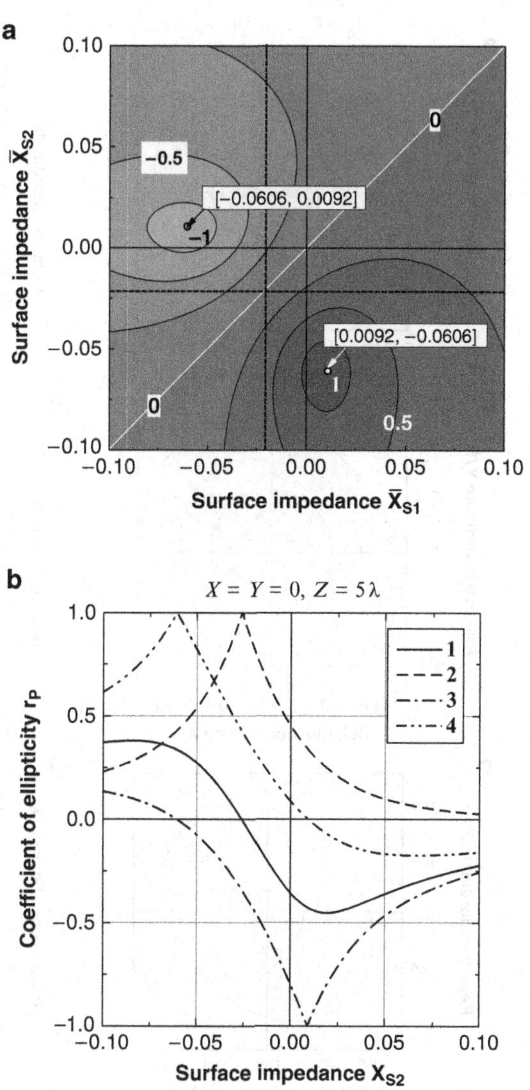

the surface impedance of each crossed vibrator. This is proved by the calculated ellipticity $r_{\mathrm{p}}(\overline{X}_{S1}, \overline{X}_{S2})$ shown in Fig. 3.14a as a function of the impedance for each vibrator for $h = \lambda/4$, $2L_{1,2} = 2L = \lambda/2$, $\Delta\Psi = 0,\ \pi/2$. Figure 3.14b represents the ellipticity r_{p} as a function of the second vibrator's impedance for various impedances of the first vibrator $\overline{X}_{S1}$ and the phase shift $\Delta\Psi$ in the feeder.

The contour plots for the values of $|\bar{E}|^2$ in the range from 0 to -30 dB with a step size of 5 dB for a system of crossed vibrators in the planes $Z = \text{const}$ are shown in Fig. 3.15. Here and below, the bar over a symbol designates normalization by

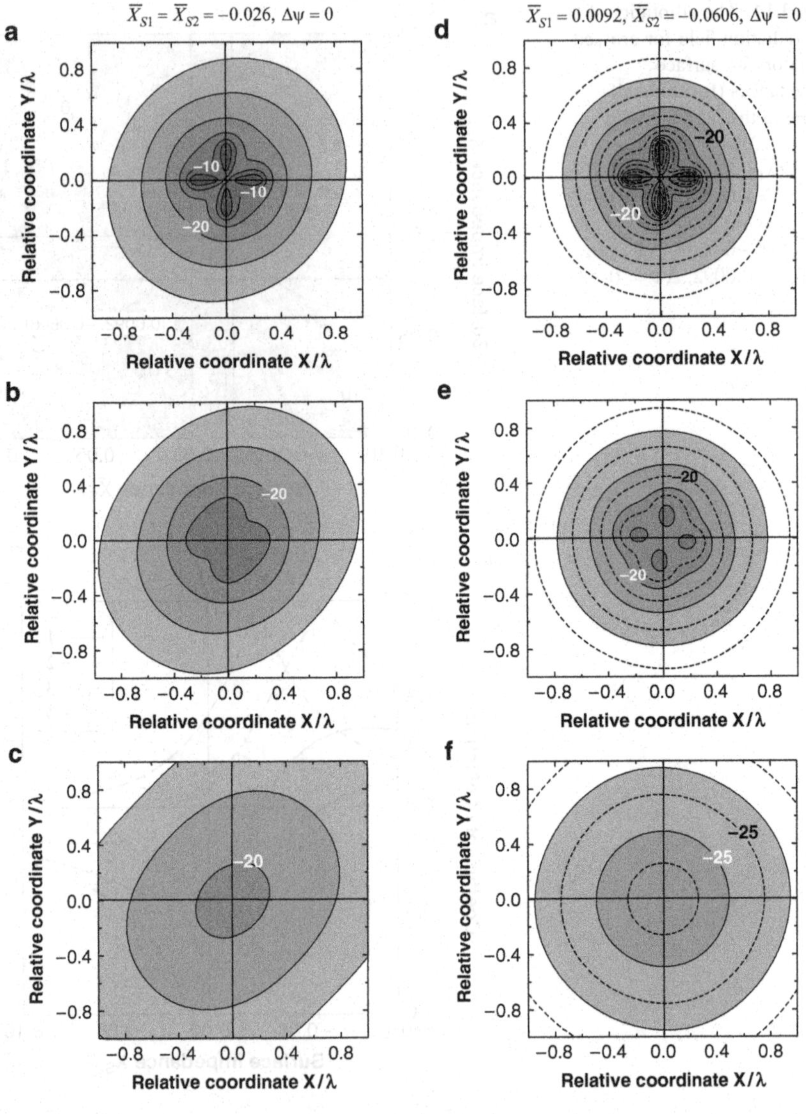

Fig. 3.15 The contour plots of $|\overline{E}|^2$ for crossed half-wave vibrators in free half-space at $Z = $ const: (**a**, **d**), $Z/\lambda = 0.025$; (**b**, **e**), $Z/\lambda = 0.1$; (**c**, **f**), $Z/\lambda = 0.5$. (**a–c**) $\overline{X}_{S1} = \overline{X}_{S2} = -0.026$, $\Delta\Psi = 0$. (**d–f**) $\overline{X}_{S1} = 0.0092$, $\overline{X}_{S2} = -0.0606$, $\Delta\Psi = 0$

the maximal value of $|\vec{E}|^2_{max}$. All plots are normalized by the maximal value of the radiation power for the resonant crossed vibrators with phase shift $\Delta\Psi = \pi/2$ (the dotted lines in Fig. 3.15d–f) in the plane $Z = 0.025\lambda$. The contour lines for the resonant vibrators (Fig. 3.15a–c) are ellipses and correspond to the linear polarization. Figure 3.15d–f shows that the contour plots are similar for both methods of circular polarization formation, and the difference in absolute values is about 3 dB. Thus, we can conclude that the formation of radiation fields with circular polarization by a system of crossed vibrators with different surface impedances is quite effective.

The contour plots of the ellipticity for the system of quarter-wave vibrators in different medium for $\rho = 5\lambda_1$, $\theta = \varphi = \pi/2$ are presented in Fig. 3.16a, c, e.

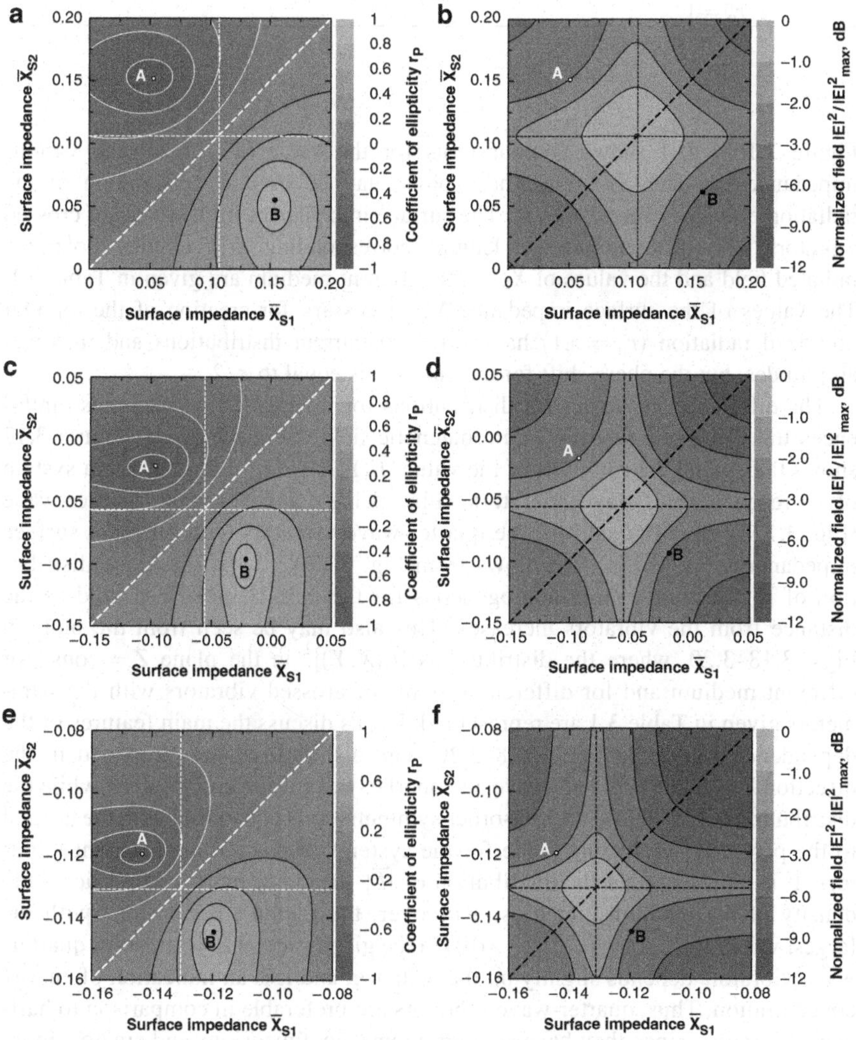

Fig. 3.16 The ellipticity r_p and $|\overline{E}|^2$ vs. the surface impedances for a system of the crossed quarter-wave vibrators ($\rho = 5\lambda_1$): (**a**, **b**), free half-space; (**c**, **d**), fat layer; (**e**, **f**), muscular tissue

Table 3.1 The problem parameters and the surface impedance $\overline{X}_{Sn}$

Parameter			Free half-space	Fat layer	Muscular tissue
λ (cm)				10	
ε_1			1.0	6.5 − i1.6	46.5 − i18.0
λ_1 (cm)			10.0	2.92	1.44
$\overline{R}_{Sn}$				0.001	
h (cm)			2.5 ($0.25\lambda_1$)	1.0 ($\approx 0.25\lambda_1$)	1.1 ($\approx 0.75\lambda_1$)
$\{\overline{X}_{S1;2}\}$	Resonant vibrators	$2L = \lambda/2$	{−0.026;−0.026}	{−0.131;−0.131}	{−0.18;−0.18}
		$2L = \lambda/4$	{0.106;0.106}	{−0.055;−0.055}	{−0.131;−0.131}
	Circular polarization $r_p = 1$	$2L = \lambda/2$	{0.009;−0.061}	{−0.116;−0.147}	{−0.174;−0.188}
		$2L = \lambda/4$	{0.152;0.052}	{−0.098;−0.022}	{−0.120;−0.145}

Figure 3.16b, d, f shows contour plots for the value of $|\overline{E}|^2$ for the various impedances of each vibrator. These plots may be used for estimation of the radiation efficiency for a given r_p. The surface impedances for a system of crossed vibrators with different electric lengths corresponding to a circular polarized radiated field and the values of $\overline{X}_{S\,\mathrm{res}}$ for different medium are given in Table 3.1. The values of the surface impedance $\overline{X}_{Sn}$ necessary for creation of the circular polarized radiation ($r_p = \pm 1$) have different current distributions and maximal amplitudes, but the phase shift for the currents is equal to $\pi/2$.

The difference in the current distributions for the crossed vibrators also influences the distribution of the electromagnetic field energy in space. Figure 3.17 shows the spatial distributions of the value $|\overline{E}(Y,Z)|^2$ ($\varepsilon_1 = \mu_1 = 1$) for a system of a resonant half-wave vibrators (Fig. 3.17a) and resonant quarter-wave (Fig. 3.17b) vibrators, and for the quarter-wave vibrators with different surface impedances (Fig. 3.17c). It follows from Fig. 3.17b, c, that the space distribution of $|\overline{E}|^2$ becomes more homogeneous for the circular polarized field as the distance from the vibrators increases. This also may be seen from the plots in Figs. 3.18–3.20, where the distributions $|\overline{E}(X,Y)|^2$ in the plane $Z = \mathrm{const}$ for different medium and for different systems of crossed vibrators with the parameters given in Table 3.1 are represented. Let us discuss the main features of the dependencies given in Figs. 3.18–3.20. The distributions are elongated in the direction $X = Y$ for both half-wave and quarter-wave resonant vibrators, while the distributions of $|\overline{E}|^2$ have a fourth-order symmetry axis coinciding with the normal to the perfectly conducting plane for the systems of crossed nonresonant vibrators. It is also seen that the distribution of $|\overline{E}|^2$ becomes more symmetric as the density of the medium increases. Moreover, the region's dimensions with the largest value of $|\overline{E}|^2$ (say $|\overline{E}|^2 \geq -5$ dB) in the given section $Z = \mathrm{const}$ for quarter-wave vibrators depends slightly on the distance, and it is an indication of energy concentration. Thus, quarter-wave vibrators are preferable in comparison to half-wave vibrators, since they have smaller geometric dimensions and create a more concentrated distribution of the radiation field to a greater extent, especially in denser medium.

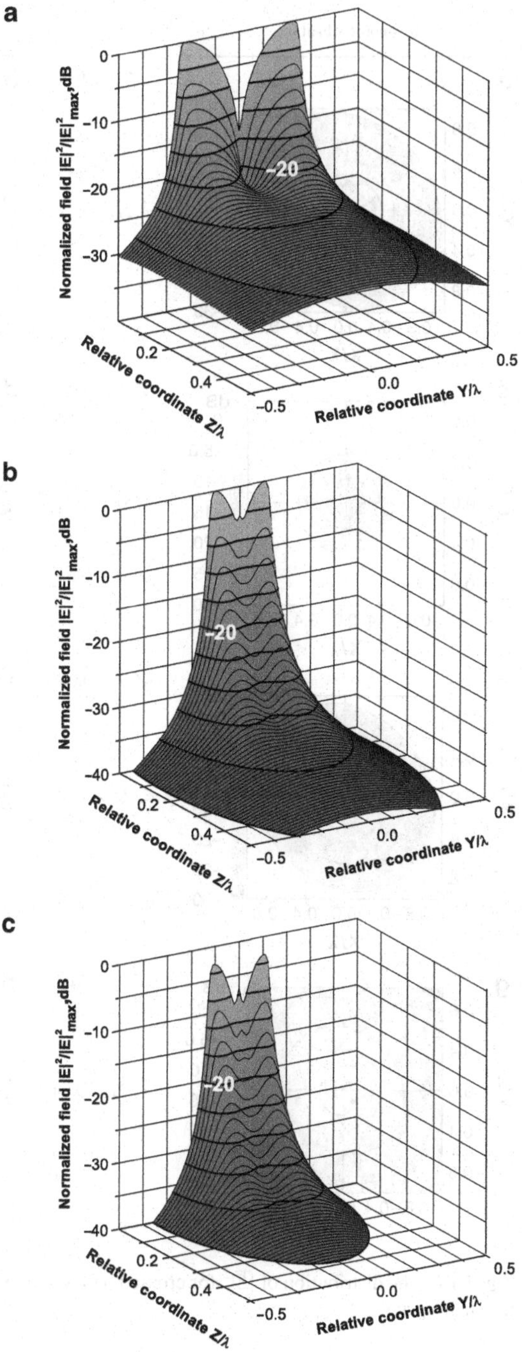

Fig. 3.17 The distribution of $|\overline{E}|^2$ in the near zone for a system of crossed vibrators in free half-space: (**a**) $2L = \lambda/2$, $\overline{X}_{S1} = \overline{X}_{S2} = -0.026$; (**b**) $2L = \lambda/4$, $\overline{X}_{S1} = \overline{X}_{S2} = 0.106$; (**c**) $2L = \lambda/4$, $\overline{X}_{S1} = 0.152$, $\overline{X}_{S2} = 0.052$

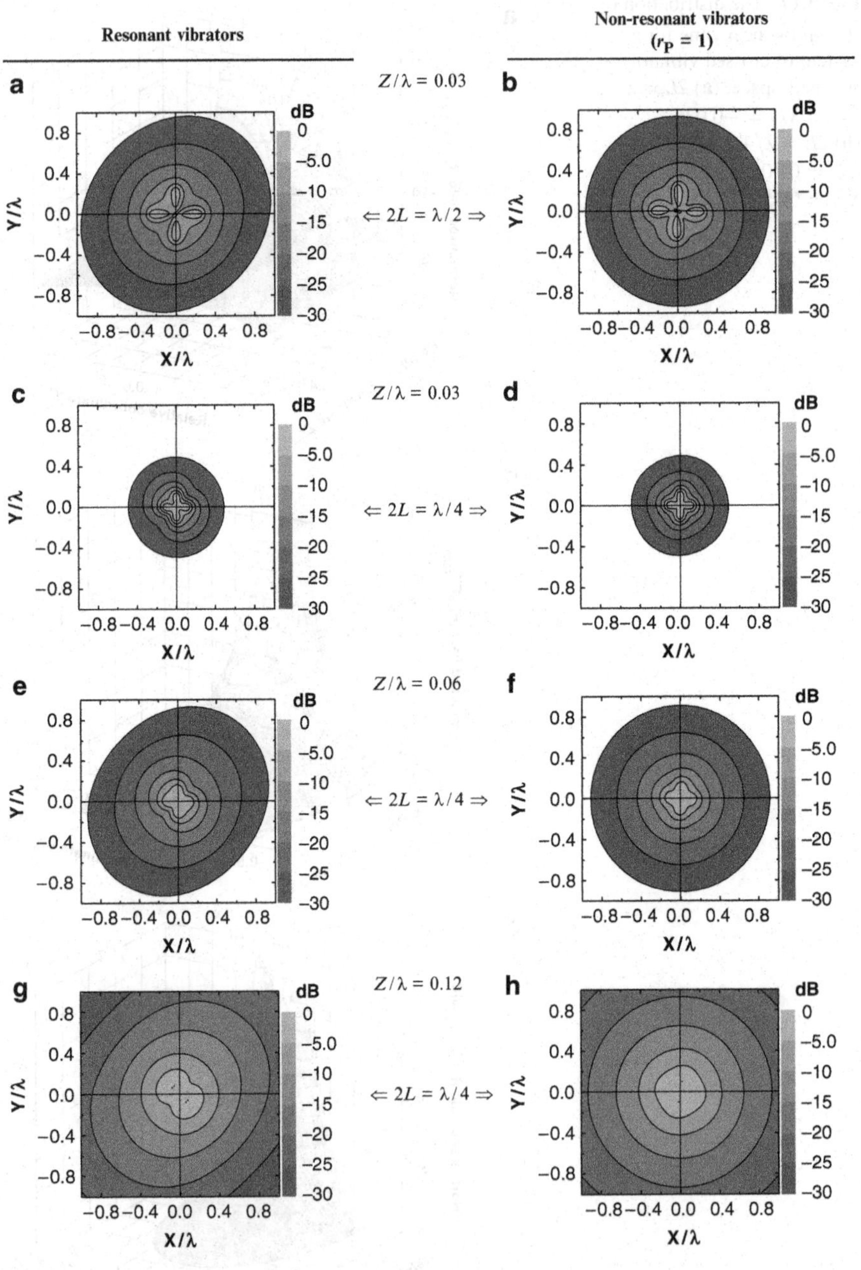

Fig. 3.18 The distribution of $|\overline{E}|^2$ for crossed half-wave (**a**, **b**) and quarter-wave (**c–h**) vibrators in free half-space

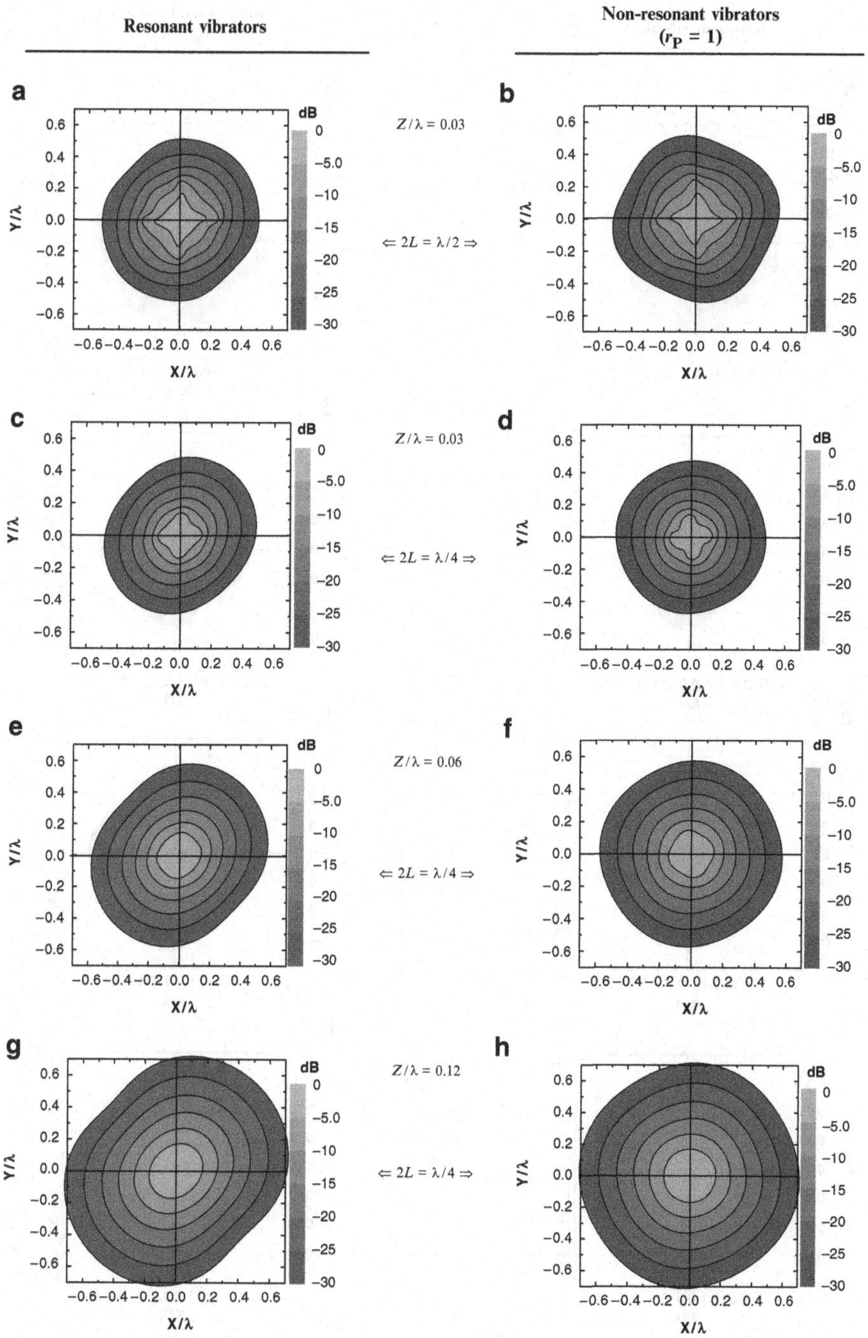

Fig. 3.19 The distribution of $|\overline{E}|^2$ for crossed half-wave (**a, b**) and quarter-wave (**c–h**) vibrators in the fat layer

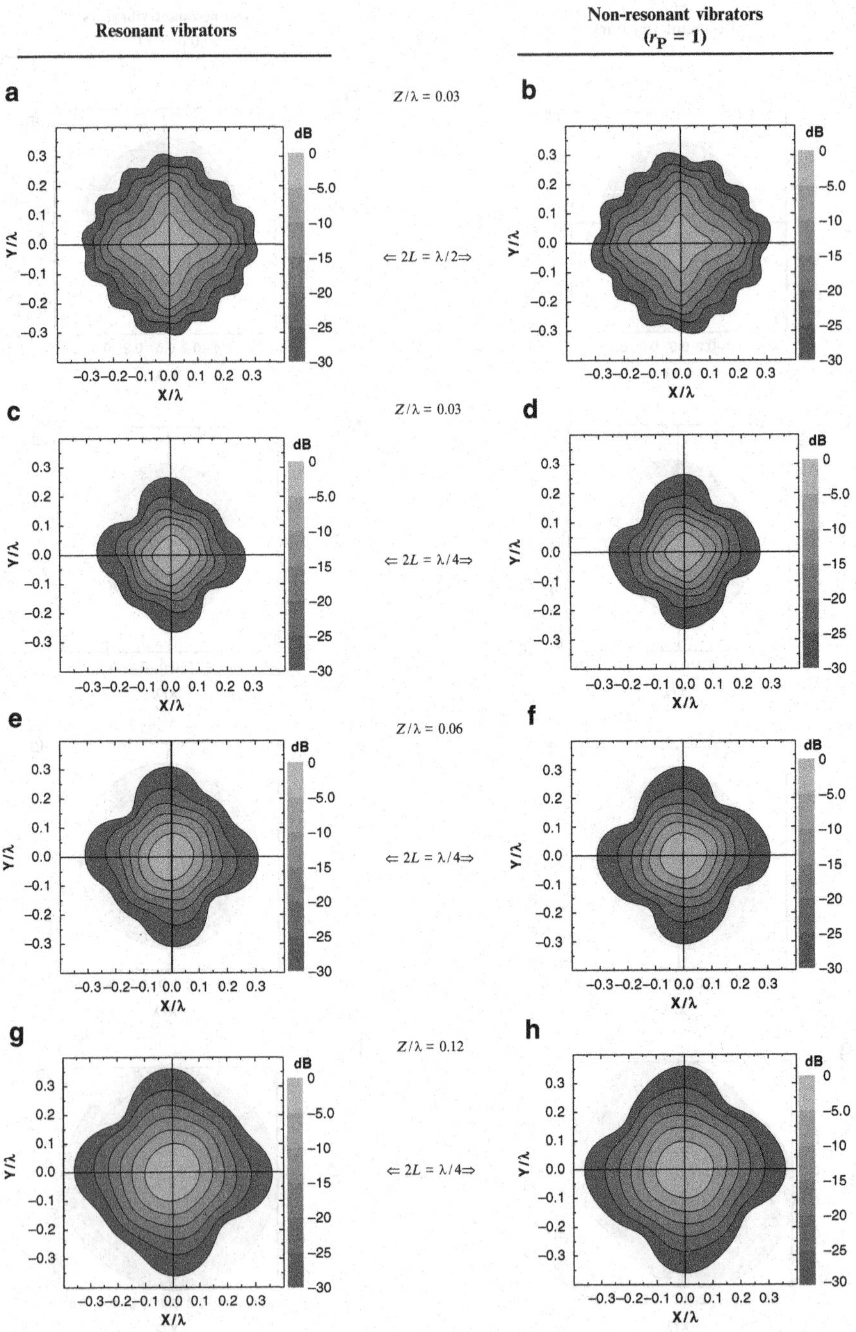

Fig. 3.20 The distribution of $|\overline{E}|^2$ for crossed half-wave (**a**, **b**) and quarter-wave (**c**–**h**) vibrators in muscular tissue

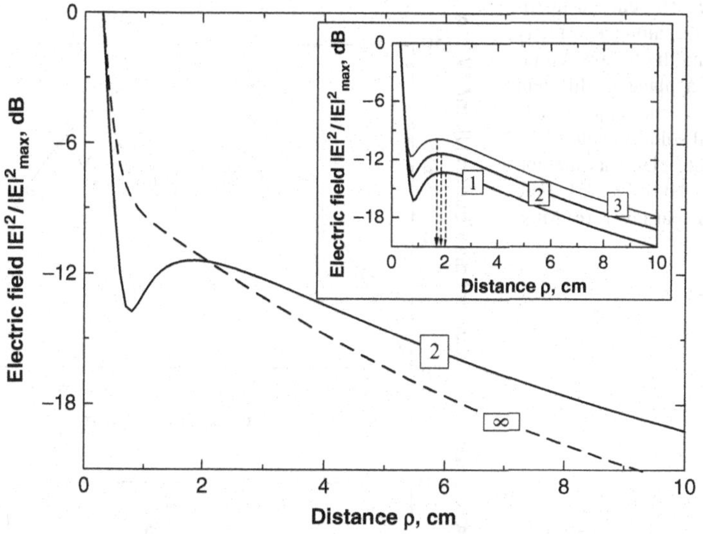

Fig. 3.21 The value of $|\bar{E}|^2$ vs. distance ρ for a system of half-wave vibrators over a plane ($\overline{X}_{S1} = -0.116, \overline{X}_{S2} = -0.147$) in the fat layer: $1 - h = 0.8$ cm, $2 - h = 1.0$ cm, $3 - h = 1.2$ cm; the dotted line corresponds to $h \to \infty$

Another feature of field formation by a system of impedance vibrators in a material medium, stipulated by the influence of a perfectly conducting screen, will be discussed below. Figure 3.21 shows the dependencies of $|\bar{E}(\rho, (\pi/2), (\pi/2))|^2$, normalized by the value for $\rho = 0.2$, for a system of crossed nonresonant half-wave vibrators. The curve in the material medium (marked as 2) decreases sharply for small ρ, reaches a minimum, and subsequently a maximum, and decreases gradually when the distance is increased. The trend of the curve differs considerably when the perfectly conducting plane is absent (the curve marked by the symbol ∞). The plot fragment in Fig. 3.21 shows the influence of the distance h between the conducting plane and the system of vibrators. The minima and maxima of the curves (shown by arrows) are shifted to the smaller h values, and the values of $|\bar{E}|^2$ increase when h is increased. This offers the possibility of a smooth tuning of the structure.

3.2.1 Comparison of Numeric Calculations Obtained by Analytical Solution and the Finite Elements Method

For a verification of the approximate analytical expression (3.2), the distribution of current amplitude along a perfectly conducting ($\overline{Z}_S = 0$) half-wave ($2L = \lambda/2$) vibrator located in different medium over a perfectly conducting plane was calculated and is shown in Fig. 3.22 as solid curves. The numerical results

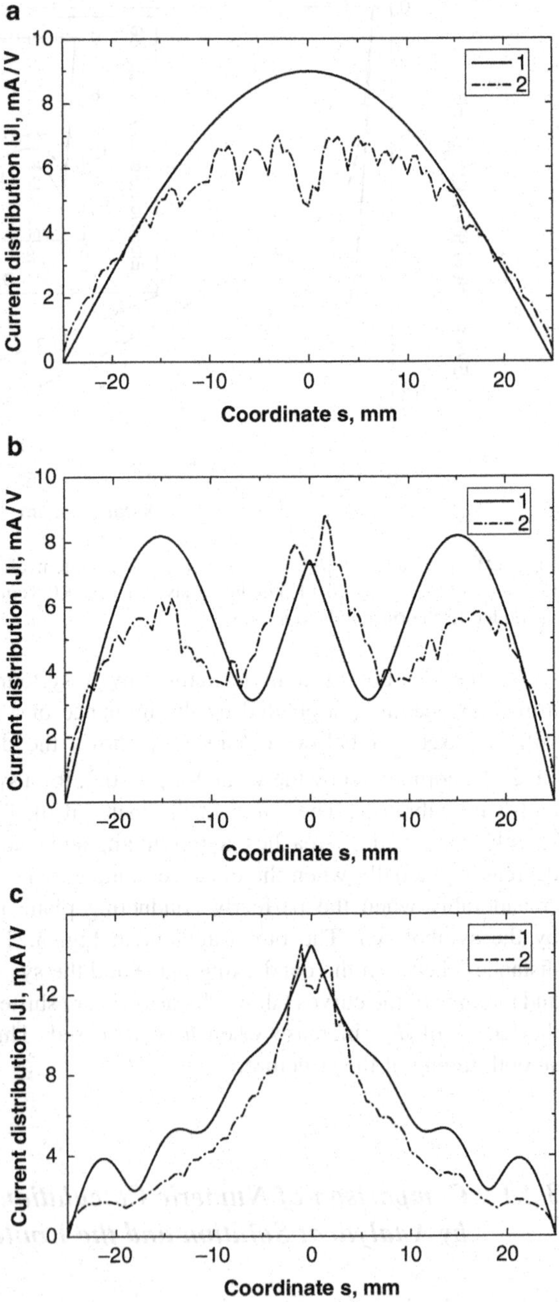

Fig. 3.22 The current distribution $|J(s)|$ along a perfectly conducting half-wave vibrator over a plane in different medium *1* the approximate analytical solution with (3.2); *2* the finite elements method ("Ansoft HFSS"). (**a**) Free space. (**b**) Fat layer. (**c**) Muscular tissue

(the dash-dotted curves) obtained in the "Ansoft HFSS" environment by the finite elements method with discretization $\Delta \approx 0.05$ are plotted here. Figure 3.22 shows that the current distributions calculated by these two quite different methods agree well as far as the curves' trends and the absolute values are concerned. A similar situation is observed for other values of the geometric parameters h and $2L$.

We have also calculated the spatial field distribution for a system of two crossed vibrators for $2L = \lambda/2$, excited with the phase shift $\Delta\Psi = \pi/2$ (Fig. 3.12). Figures 3.23–3.26 show the distributions of the normalized electrical field $|E|/|E|_{\max}$ in the two planes $Z = 6$ mm and $Z = 10$ mm, parallel to the perfectly conducting surface. For clarity, the distributions are shown as three-dimensional surface plots (the fragments (a) and (b)) and contour plots (the fragments (c) and (d)). The plots in Figs. 3.23–3.26 show that the analytical expressions (3.3) and (3.2) make it possible to calculate the spatial distributions of the vibrator fields in various medium and the differences between these results and those obtained by different methods become less for denser medium.

It should be noted that while the calculational results of different methods are well correlated, but their computational efficiencies vary considerably. Thus, the

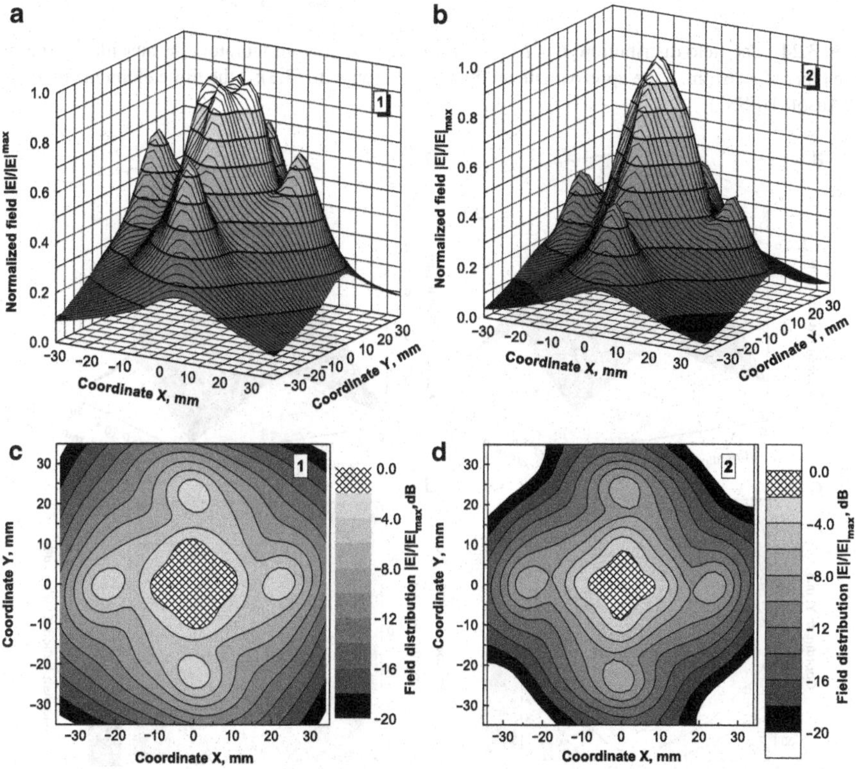

Fig. 3.23 The field distribution $|E|/|E|_{\max}$ for a system of crossed vibrators in the fat layer in the plane $Z = 6$ mm: *1* approximate analytical solution with (3.3); *2* the finite elements method ("Ansoft HFSS")

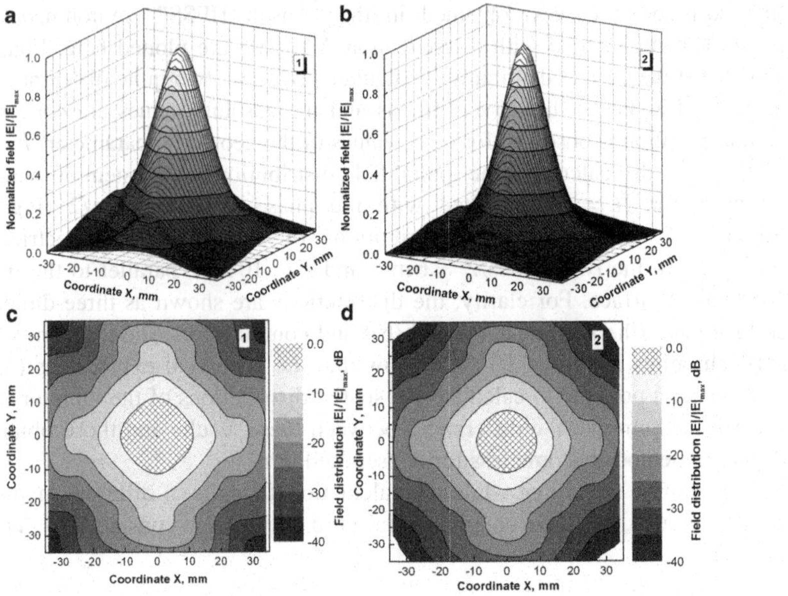

Fig. 3.24 The field distribution $|E|/|E|_{\max}$ for a system of crossed vibrators in muscular tissue in the plane $Z = 6$ mm: *1* the approximate analytical solution with (3.3); *2* the finite elements method ("Ansoft HFSS")

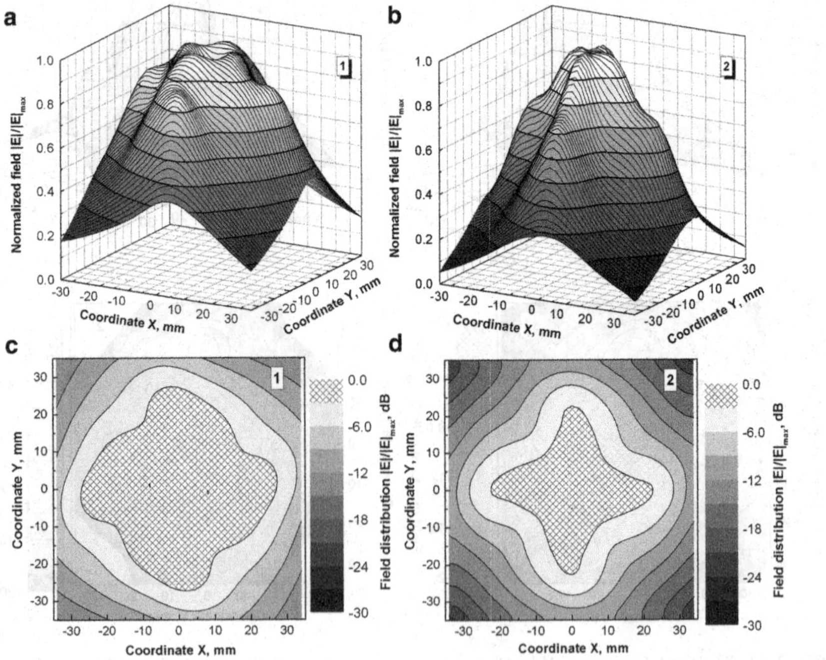

Fig. 3.25 The field distribution $|E|/|E|_{\max}$ of a system of crossed vibrators in the fat layer at $Z = 10$ mm: *1* approximate analytical solution with (3.3); *2* the finite elements method ("Ansoft HFSS")

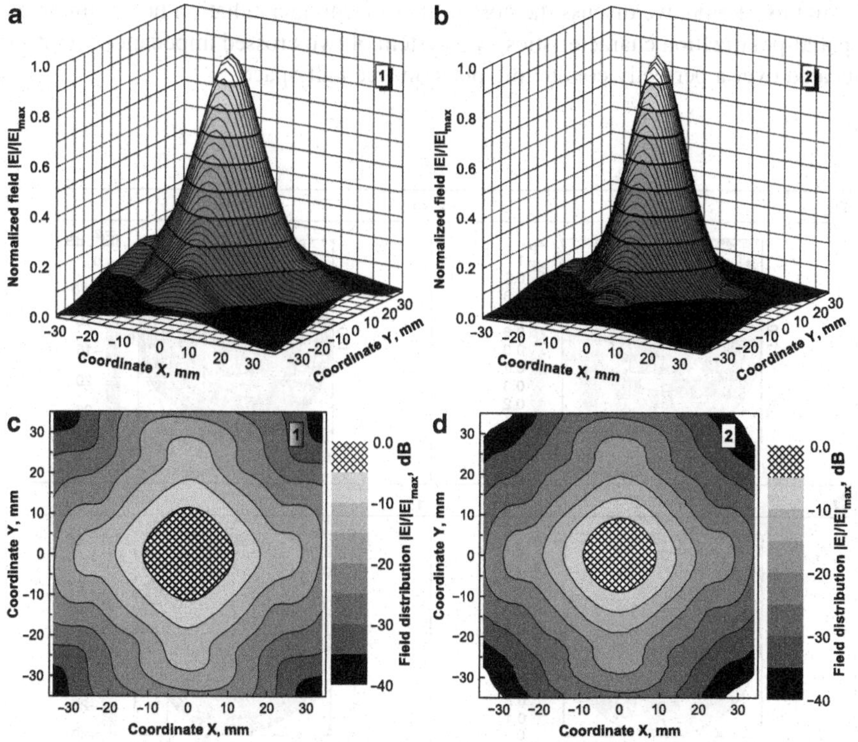

Fig. 3.26 The field distribution $|E|/|E|_{max}$ for a system of crossed vibrators in muscular tissue in the plane $Z = 10$ mm: *1* approximate analytical solution with (3.3); *2* finite elements method ("Ansoft HFSS")

running time for a PC with Intel Pentium Corel Duo 2.2 GHz for numeric computation of the current distribution in Fig. 3.22a by the finite elements method is 39 min, while an approximate analytical solution in the MathCad environment is about 2 s (15 s for the field spatial distribution).

3.3 Formation of the Radiation Field with Specified Spatial-Polarization Characteristics by a System of Crossed Impedance Vibrators

At present, elliptical (circular) polarization is widely used in antenna techniques [18]. Radio channels with elliptical polarization are applied in communication and data transfer facilities, radar, navigation, and data interception systems, etc. In practice, it is necessary that antenna radiated elliptically polarized electromagnetic field with predetermined unit ellipticity in a prescribed direction [19].

In this section, we discuss the possibility of forming a radiation field with given spatial-polarization characteristics by a system of two crossed impedance vibrators located over a perfectly conducting plane in free half-space.

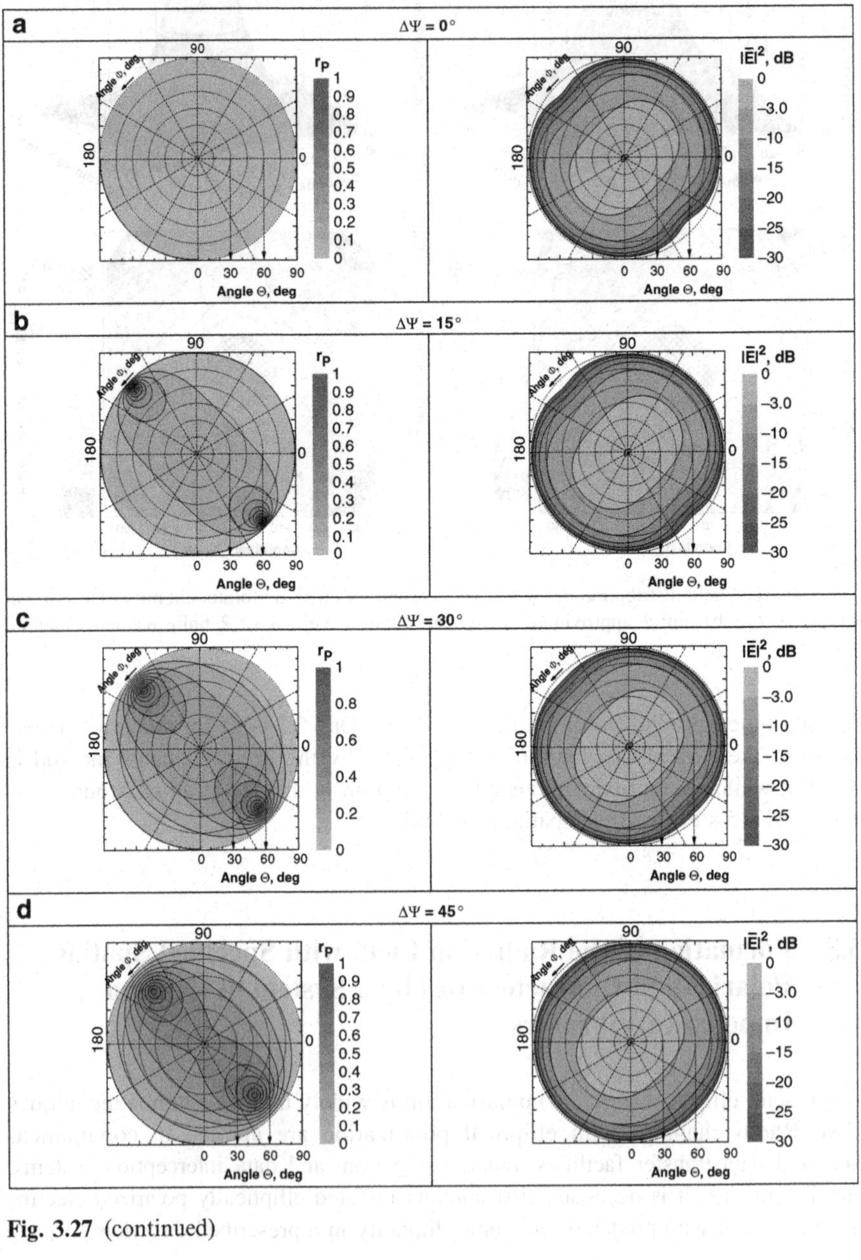

Fig. 3.27 (continued)

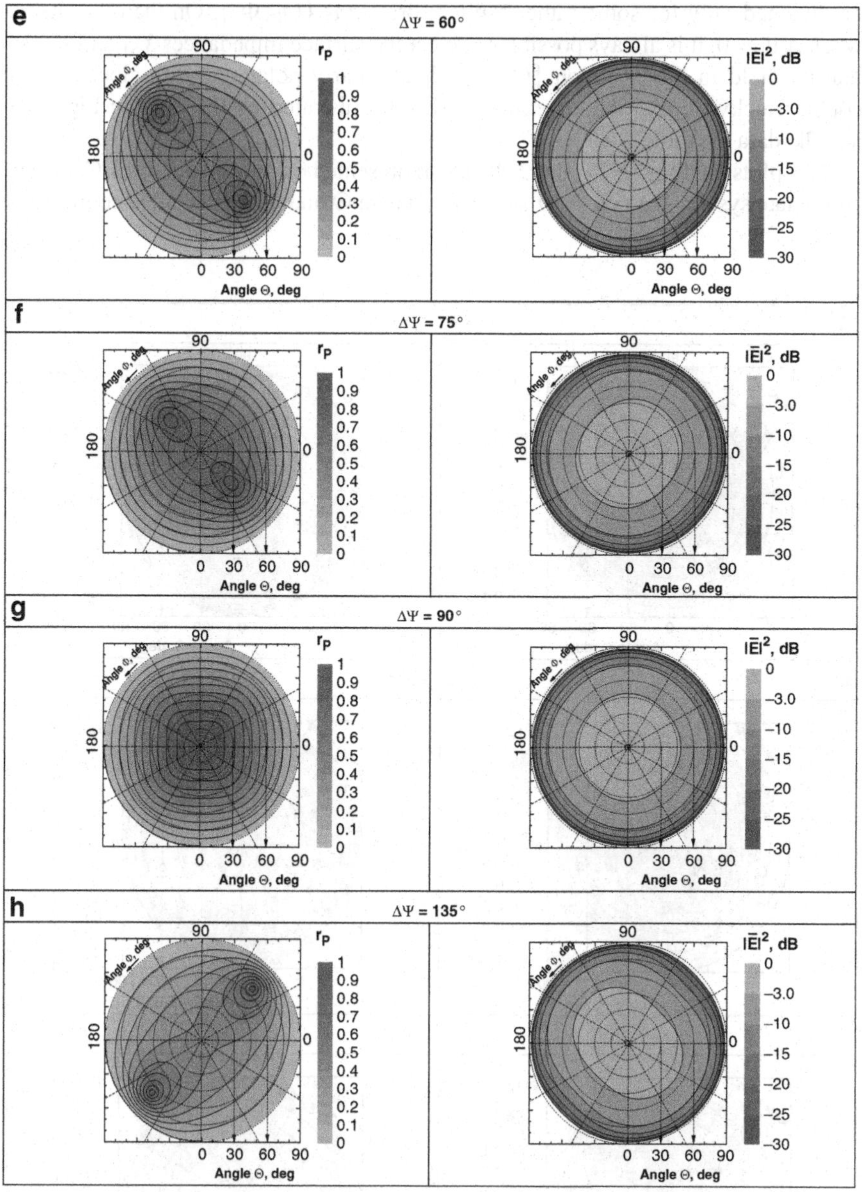

Fig. 3.27 The spatial distributions of r_{p} and $|\overline{E}|^2$ for a system of crossed perfectly conducting vibrators for different phase shifts $\Delta\Psi$. (**a**) $\Delta\Psi = 0°$. (**b**) $\Delta\Psi = 15°$. (**c**) $\Delta\Psi = 30°$. (**d**) $\Delta\Psi = 45°$. (**e**) $\Delta\Psi = 60°$. (**f**) $\Delta\Psi = 75°$. (**g**) $\Delta\Psi = 90°$. (**h**) $\Delta\Psi = 135°$

Figure 3.27 shows the spatial distributions of ellipticity r_{p} and radiation field $|\bar{E}|^2$ for a system of crossed perfectly conducting half-wave vibrators with different phase shifts $\Delta\Psi$ in their feeding tracts. As can be seen, the ellipticity $r_{\mathrm{p}} = 1.0$ may be obtained only for some rather limited directions $\{\Theta_0, \Phi_0\}$. On the other hand, when $\Delta\Psi = 0$, it is always possible to select the surface impedances $\overline{X}_{S1}$ and $\overline{X}_{S2}$ so that the field in the direction $\{\Theta_0, \Phi_0\}$ is circularly polarized with clockwise or counterclockwise rotation as required. This is illustrated by the plots in Fig. 3.28 and the data in Table 3.2.

The plots in Figs. 3.27 and 3.28 are represented in the generalized Cartesian coordinate system $\{\Theta \cos \Phi, \Theta \sin \Phi, Z\}$, where Θ and Φ are the angle coordinates

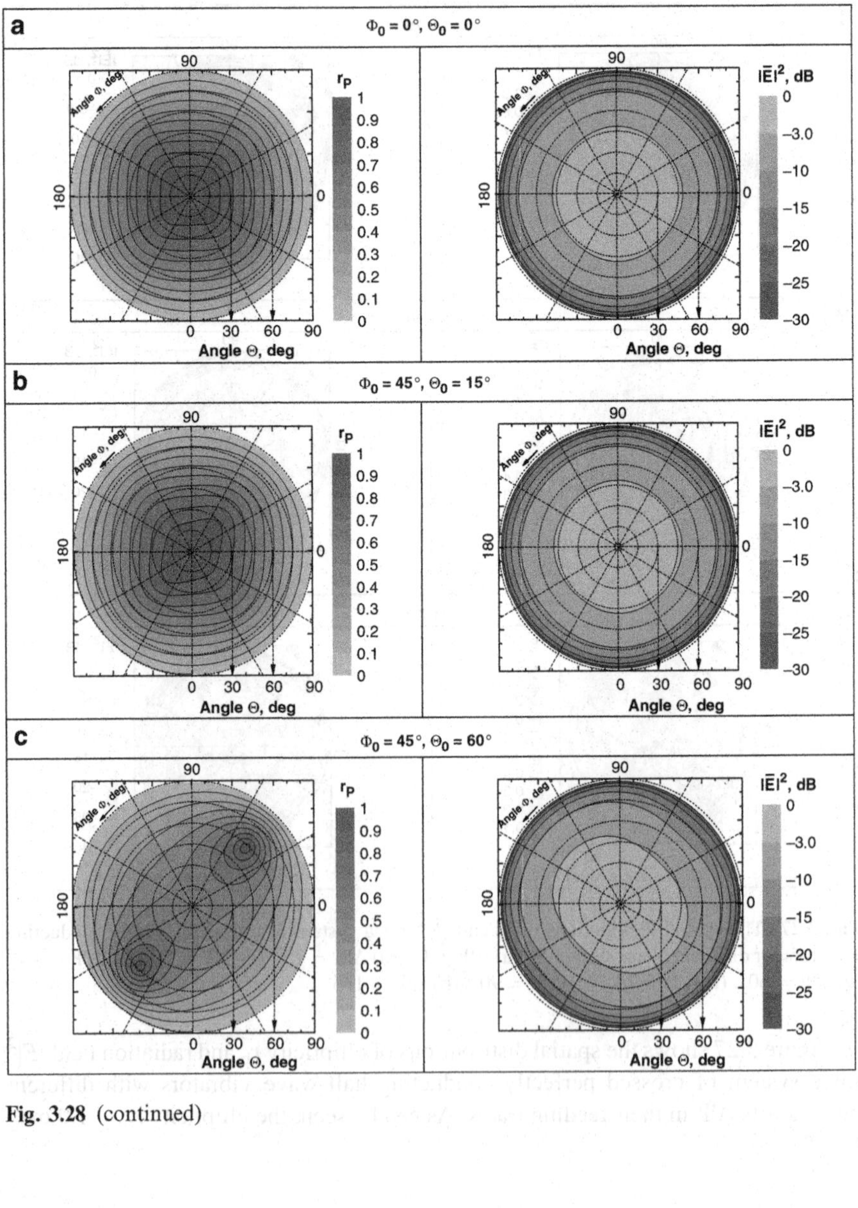

Fig. 3.28 (continued)

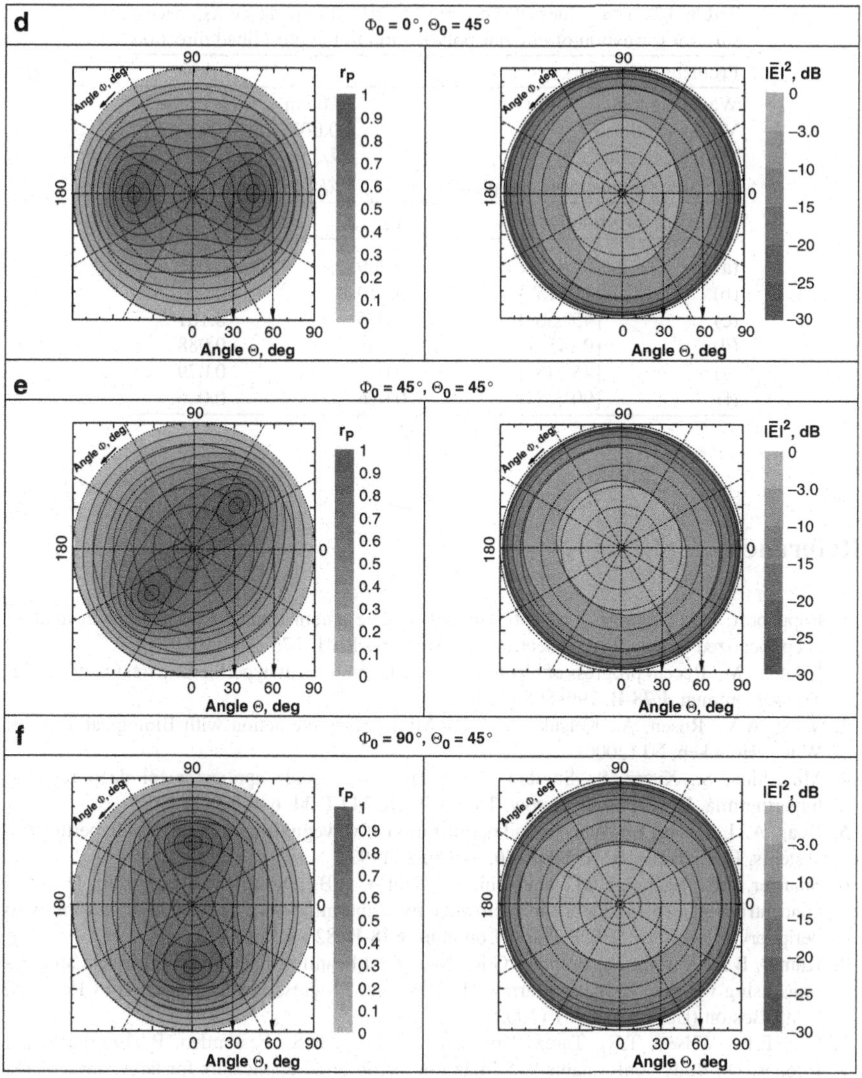

Fig. 3.28 The spatial distributions of r_p and $\overline{E}$ for a system of crossed perfectly conducting vibrators for the directions Φ_0, Θ_0. **(a)** $\Phi_0 = 0°$, $\Theta_0 = 0°$. **(b)** $\Phi_0 = 45°$, $\Theta_0 = 15°$. **(c)** $\Phi_0 = 45°$, $\Theta_0 = 60°$. **(d)** $\Phi_0 = 0°$, $\Theta_0 = 45°$. **(e)** $\Phi_0 = 45°$, $\Theta_0 = 45°$. **(f)** $\Phi_0 = 90°$, $\Theta_0 = 45°$

of the spherical coordinate system related to the Cartesian system $\{X, Y, Z\}$ as shown in Fig. 3.12. In Figs. 3.27 and 3.28, the concentric circles correspond to $\Theta = \text{const}$, and the radial rays to $\Phi = \text{const}$.

The analysis carried out in this section may be also valid for a system of crossed vibrators in a material semi-infinite medium.

Table 3.2 The values of the vibrator surface impedance $\overline{X}_{Sn}$ necessary for formation of circular polarization in the prescribed direction

Problem parameters			
Wavelength λ		10 cm	
Vibrator radius r		$0.0033\lambda = 0.033$ cm	
Vibrator length $2L$		$\lambda/2 = 5$ cm	
Vibrator distance from the plane h		$\lambda/4 = 2.5$ cm	
$\{\Phi_0, \Theta_0\}$ direction for $r_p = 1$		$\overline{X}_{S1}$	$\overline{X}_{S2}$
(a)	{0°,0°}	0.009	−0.061
(b)	{45°,15°}	0.010	−0.062
(c)	{45°,60°}	0.035	−0.101
(d)	{0°,45°}	−0.002	−0.088
(e)	{45°,45°}	0.021	−0.079
(f)	{90°,45°}	0.026	−0.046

References

1. Rappaport, C.M.: Synthesis of optimum microwave antenna applicators for use in treating deep localized tumors. Prog. Electromagn. Res. **PIER 01**, 175–240 (1989).
2. Kikuchi, M.: Recent progress of electromagnetic techniques in hyperthermia treatment. IEICE Trans. Commun. **E78-B**, 799–808 (1995).
3. Vorst, A.V., Rosen, A., Kotsuka, Yu.: RF/Microwave Interaction with Biological Tissues. Wiley, Hoboken, NJ (2006).
4. Mizushina, S., Xiang, S., Sugiura, T.: A large waveguide applicator for deep regional hyperthermia. IEEE Trans. Microw. Theory Tech. **MTT-34**, 644–648 (1986).
5. Boag, A., Leviatan, Y.: Analysis and optimization of waveguide multiapplicator hyperthermia systems. IEEE Trans. BME **BME-40**, 946–952 (1993).
6. Stauffer, P.R., Leoncini, M., Manfrini, V., Gentili, G.B., Diederich, C.J., Bozzo, D.: Dual concentric conductor radiator for microwave hyperthermia with improved field uniformity to periphery of aperture. IEICE Trans. Commun. **E78-B**, 826–835 (1995).
7. Kumar, B.P., Karnik, N., Branner, G.R.: Near-field beamforming for hyperthermia applications using waveguide aperture arrays. In Proc. Int. Progress in Electromagnetics Research Conf. Boston (USA), 345–351 (2002).
8. Lee, E.R., Wilsey, T.R., Tarczy-Hornoch, P., Kapp, D.S., Fessenden, P., Lohrbach, A., Prionas, D.: Body conformable 915 MHz microstrip array applicators for large surface area hyperthermia. IEEE Trans. BME **BME-39**, 470–483 (1992).
9. Shimotori, T., Nikawa, Y., Mori, S.: Study on semicylindrical microstrip applicator for microwave hyperthermia. IEICE Trans. Electron. **E77-C**, 942–948 (1994).
10. Nikawa, Y., Yamamoto, M.: A multielement flexible microstrip patch applicator for microwave hyperthermia. IEICE Trans. Commun. **E78-B**, 145–151 (1995).
11. Cresson, P.-Y., Michel, C., Dubois, L., Chive, M., Pribetich, J.: Complete three-dimensional modeling of new microstrip-microslot applicators for microwave hyperthermia using the FDTD method. IEEE Trans. Microw. Theory Tech. **MTT-42**, 2657–2666 (1994).
12. Tanaka, R., Nikawa, Y., Mori, S.: A dielectric rod waveguide applicator for microwave hyperthermia. IEICE Trans. Commun. **E76-B**, 703–798 (1993).
13. Hamada, L., Wu, M.-S., Ito, K., Kasai, H.: Basic analysis on SAR distribution of coaxial-slot antenna array for interstitial microwave hyperthermia. IEICE Trans. Electron. **E78-C**, 1624–1631 (1995).

14. Hamada, L., Yoshimura, H., Ito, K.: A new feeding technique for temperature distribution control in interstitial microwave hyperthermia. IEICE Trans. Electron. **E82-C**, 1318–1323 (1999).
15. Saito, K., Yoshimura, H., Ito, K.: Numerical simulation for interstitial heating of actual neck tumor based on MRI tomograms by using a coaxial-slot antenna. IEICE Trans. Electron. **E86-C**, 2482–2487 (2003).
16. Wu, M.-S., Hamada, L., Ito, K., Kasai, H.: Effect of a catheter on SAR distribution around interstitial antenna for microwave hyperthermia. IEICE Trans. Commun. **E78-B**, 845–850 (1995).
17. Saito, K., Taniguchi, T., Yoshimura, H., Ito, K.: Estimation of SAR distribution of a tip-split array applicator for microwave coagulation therapy using the finite element method. IEICE Trans. Electron. **E84-C**, 948–954 (2001).
18. Skolnik, M.I. (editor-in-chief): Radar Handbook. McGraw-Hill, New York (1970).
19. Zemlyanskiy, S.V., Mishchenko, S.E., Shatskiy, V.V.: A weakly directional antenna with the controlled ellipticity coefficient. In Proc. VI International Symposium on Electromagnetic Compatibility and Electromagnetic Ecology. Saint-Petersburg (Russia), 135–137 (2005).

Chapter 4
Electromagnetic Waves Scattering by Irregular Impedance Vibrators in Free Space

Variation of the cross section and (or) surface impedance along a vibrator may serve as additional parameters for the formation of preassigned electrodynamic characteristics of cylindrical vibrator antennas. For example, the broadbandedness of a biconical vibrator whose arms are two cones with vertices directed to the center is much larger than that of a vibrator with constant radius. Since pioneering work by Schelkunoff [1], antennas of such configurations have attracted the attention of many investigators (see, for example, [2–6]). Vibrators with variable surface impedance, yielding additional possibilities for controlling the characteristics of antennas of fixed geometric size, take a special place among thin impedance vibrators [7–12]. We will not analyze the methods employed for solving problems described in the above-mentioned publications, but simply note that they are devoted to the investigation of radiating characteristics (both theoretical and experimental) of vibrators excited at the center by concentrated EMF. However, for the analysis of receiving antennas it is necessary to know the induced current in a scattering vibrator excited by an incident electromagnetic wave. This problem is also important for the investigation of scattering by material bodies having complex shapes ([18] in Chap. 3).

4.1 Impedance Vibrators with Variable Radius

In this section we derive an approximate analytical solution to the problem of scattering of electromagnetic waves by thin impedance vibrators with variable cross-sectional radius by the averaging method, and we also investigate the electrodynamic characteristics of such antennas.

We restrict the examination to the linear law of radius variation (Fig. 4.1), which may be a good approximation for other dependencies, e.g., exponential for small angles ψ. Let the impedance vibrator in free space be excited by a plane electromagnetic wave of amplitude E_0. The vibrator length is $2L$, and its variable radius

M.V. Nesterenko et al., *Thin Impedance Vibrators*, Lecture Notes in Electrical Engineering 2064, DOI 10.1007/978-1-4419-7850-9_4,

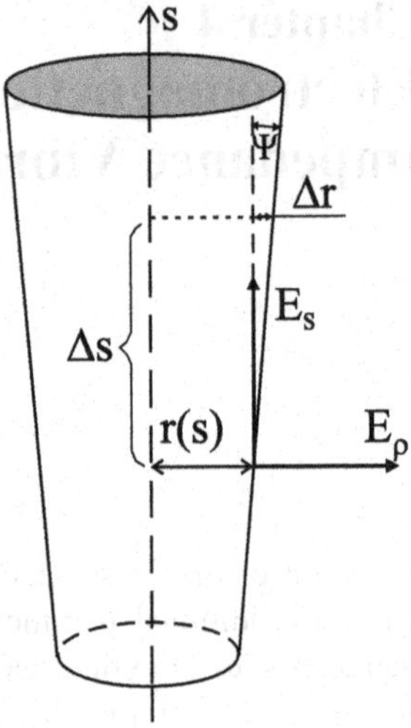

Fig. 4.1 A fragment of a vibrator

$r(s)$ is such that the vibrator is electrically thin and the following inequalities are valid:

$$kr(s) \ll 1, \; r(s) \ll 2L. \tag{4.1}$$

Then the tangential component of the scattered field on the vibrator's surface in the cylindrical coordinate system $\{\rho, \varphi, s\}$ is given by

$$E_\tau^{\,sc}(\rho, s) = E_s(\rho, s)\cos\psi + E_\rho(\rho, s)\sin\psi. \tag{4.2}$$

Here

$$E_s(\rho, s) = \frac{\partial}{\partial s}\left[\frac{1}{\rho}\frac{\partial(\rho\Pi_\rho)}{\partial\rho} + \frac{\partial\Pi_s}{\partial s}\right] + k^2\Pi_s,$$

$$E_\rho(\rho, s) = \frac{\partial}{\partial\rho}\left[\frac{1}{\rho}\frac{\partial(\rho\Pi_\rho)}{\partial\rho} + \frac{\partial\Pi_s}{\partial s}\right] + k^2\Pi_\rho,$$

and Π_s and Π_ρ are the components of the electrical Hertz vector corresponding to the vibrator current $\vec{J}(s) = \vec{e}_s J_s(s) + \vec{e}_\rho J_\rho(s)$, where $\vec{e}_s$, $\vec{e}_\rho$ are the unit vectors for the cylindrical coordinate system.

We assume $|J_\rho| \ll |J_s|, J_s \approx J$, in accordance with (4.1), and after conversion to the total derivative with respect to the longitudinal coordinate s, with regard to the relation $\mathrm{d}r(s)/\mathrm{d}s = \tan\psi$, we obtain the integrodifferential equation relative to the current $J(s)$ for the impedance boundary condition on the vibrator's surface:

$$\left(\frac{\mathrm{d}^2}{\mathrm{d}s^2}+k^2\right)\int\limits_{-L}^{L}J(s')\frac{\mathrm{e}^{-ik\tilde{R}(s,s')}}{\tilde{R}(s,s')}\mathrm{d}s'$$
$$=-\frac{\mathrm{i}\omega}{\cos\psi}E_{0\tau}(s)-\tan\psi\frac{\mathrm{d}}{\mathrm{d}s}\left\{r(s)\int\limits_{-L}^{L}J(s')\frac{\mathrm{e}^{-ik\tilde{R}(s,s')}}{\tilde{R}^3(s,s')}\mathrm{d}s'\right\}+\frac{\mathrm{i}\omega}{\cos\psi}z_iJ(s),\tag{4.3}$$

where $E_{0\tau}(s)$ is the tangential component of the electrical field induced by the impressed sources, $\tilde{R}(s,s')=\sqrt{(s-s')^2+r^2(s)}$. Equation (4.3) transits into (2.1) for the current in an impedance vibrator with constant radius $r(s)=\text{const}=r_0$.

One can show by straightforward differentiation that for $r(s)=r_0$,

$$\frac{\mathrm{d}^2}{\mathrm{d}s^2}\frac{\mathrm{e}^{-ik\tilde{R}(s,s')}}{\tilde{R}(s,s')}\approx\frac{\mathrm{d}^2}{\mathrm{d}s'^2}\frac{\mathrm{e}^{-ik\tilde{R}(s,s')}}{\tilde{R}(s,s')},\tag{4.4}$$

and the second summand on the right-hand side of (4.3) equals zero when $s'=s$, that is, it is not singular:

$$\tan\psi\frac{\mathrm{d}}{\mathrm{d}s}\left\{r(s)\int\limits_{-L}^{L}J(s')\frac{\mathrm{e}^{-ik\tilde{R}(s,s')}}{\tilde{R}^3(s,s')}\mathrm{d}s'\right\}\cong\tan\psi r(s)\int\limits_{-L}^{L}J(s')(s-s')3\frac{\mathrm{e}^{-ik\tilde{R}(s,s')}}{\tilde{R}^5(s,s')}\mathrm{d}s'\Bigg|_{s=s'}=0.$$

Let us isolate the main part of the kernel in (4.3) as in (2.2):

$$\int\limits_{-L}^{L}J(s')\frac{\mathrm{e}^{-ik\tilde{R}(s,s')}}{\tilde{R}(s,s')}\mathrm{d}s'=J(s)\,\Omega(s)+\int\limits_{-L}^{L}\frac{[J(s')\,\mathrm{e}^{-ik\tilde{R}(s,s')}-J(s)]}{\tilde{R}(s,s')}\mathrm{d}s'.\tag{4.5}$$

Here $\Omega(s)=\int\limits_{-L}^{L}\frac{\mathrm{d}s'}{\tilde{R}(s,s')}=\Omega+\tilde{\gamma}(s,r(s))$,

$$\tilde{\gamma}(s,r(s))=\ln\left\{\left(\frac{r_0}{r_L}\right)^2\frac{[(L+s)+\sqrt{(L+s)^2+r^2(s)}][(L-s)+\sqrt{(L-s)^2+r^2(s)}]}{4L^2}\right\},$$

$\Omega=2\ln(2l/r_L)\gg 1$, and r_0 and r_L are the radii of the vibrator at the its center and end, respectively. The natural large parameter Ω for angles $\psi\leqslant 10°$ to the constant multiplier coincides with the characteristic impedance of an infinite biconical antenna modeled by TEM-wave propagation in a homogeneous transmission line [1].

Using equality (4.4) and neglecting the current on the vibrator's ends ($J(\pm L)=0$) [3], we obtain the following integrodifferential equation with the small parameter α:

$$\frac{d^2J(s)}{ds^2} + k^2J(s) = \alpha\left\{\frac{i\omega}{\cos\psi}E_{0\tau}(s) + F[s,J(s)] - \frac{i\omega}{\cos\psi}\frac{Z_s}{2\pi r(s)}J(s)\right\}. \tag{4.6}$$

Here $\alpha = 1/(2\ln[r_L/(2L)])$, ($|\alpha| \ll 1$),

$$F[s,J(s)] = -\frac{dJ(s')}{ds'}\frac{e^{-ik\tilde{R}(s,s')}}{\tilde{R}(s,s')}\Bigg|_{-L}^{L} + [J''(s) + k^2J(s)]\tilde{\gamma}(s,r(s))$$
$$+\int_{-L}^{L}\frac{\{[J''(s') + k^2J(s')]\,e^{-ik\tilde{R}(s,s')} - [J''(s) + k^2J(s)]\}}{\tilde{R}(s,s')}ds'$$
$$+\tan\psi r(s)\left[\int_{-L}^{L}\frac{e^{-ik\tilde{R}(s,s')}}{\tilde{R}^3(s,s')}ds'\right]$$
$$\times\left\{\frac{dJ(s')}{ds'}\Bigg|_{-L}^{L} - \int_{-L}^{L}[J''(s) + k^2J(s)]ds\right\}. \tag{4.7}$$

In the operator $F[s, J(s)]$ defining the vibrator self-field and $J''(s)$ and $J''(s')$ denote the second current derivatives with respect to coordinates s and s'.

Let us use the asymptotic averaging method, whose efficiency was proved in Sect. 2.1, to obtain an approximate solution to (4.6). We reduce (4.6) to a standard system of equations with a small parameter by the change of variables (2.6). Then (4.6) may be converted to a system of the integrodifferential equations, unresolved for derivatives:

$$\begin{aligned}\frac{dA(s)}{ds} &= -\frac{\alpha}{k}\left\{\begin{array}{l}\dfrac{i\omega}{\cos\psi}E_{0\tau}(s) + F\left[s, A(s), \dfrac{dA(s)}{ds}, B(s), \dfrac{dB(s)}{ds}\right]\\ -\dfrac{i\omega Z_s}{2\pi\cos\psi\, r(s)}[A(s)\cos ks + B(s)\sin ks]\end{array}\right\}\sin ks,\\ \frac{dB(s)}{ds} &= +\frac{\alpha}{k}\left\{\begin{array}{l}\dfrac{i\omega}{\cos\psi}E_{0\tau}(s) + F\left[s, A(s), \dfrac{dA(s)}{ds}, B(s), \dfrac{dB(s)}{ds}\right]\\ -\dfrac{i\omega Z_s}{2\pi\cos\psi\, r(s)}[A(s)\cos ks + B(s)\sin ks]\end{array}\right\}\cos ks.\end{aligned} \tag{4.8}$$

Carrying out partial averaging in the variable s and using (4.1) in (4.8), we obtain the system of ordinary differential equations for the first approximation relative to the averaged functions $\overline{A}(s)$ and $\overline{B}(s)$:

$$\begin{aligned}\frac{d\overline{A}(s)}{ds} &= -\alpha\left\{\frac{i\omega}{k\cos\psi}E_{0\tau}(s) + \overline{F}[s,\overline{A}(s),\overline{B}(s)]\right\}\sin ks + \chi_v\overline{B}(s),\\ \frac{d\overline{B}(s)}{ds} &= +\alpha\left\{\frac{i\omega}{k\cos\psi}E_{0\tau}(s) + \overline{F}[s,\overline{A}(s),\overline{B}(s)]\right\}\cos ks - \chi_v\overline{A}(s).\end{aligned} \tag{4.9}$$

Here $\chi_v = (\mathrm{i}\alpha/r_\mathrm{L}\cos\psi)(3/2 - r_0/(2r_\mathrm{L}))\overline{Z}_\mathrm{S}$. For $r_L = r_0$, the expression for χ_v reduces to $\chi = \mathrm{i}(\alpha\overline{Z}_\mathrm{S}/r_0)$, which accounts for the surface impedance of a vibrator with constant radius r_0, and

$$\overline{F}[s,\overline{A}(s),\overline{B}(s)] = [\overline{A}(s')\sin ks' - \overline{B}(s')\cos ks']\frac{\mathrm{e}^{-\mathrm{i}k\tilde{R}(s,s')}}{\tilde{R}(s,s')}\Bigg|_{-L}^{L} \qquad (4.10)$$

is the vibrator self-field, averaged along its length.

Integrating (4.9) and taking into account (4.10), we obtain an asymptotic expression for the current in an impedance vibrator as

$$J(s) = \overline{A}(-L)\cos(\tilde{k}s + \chi_v L)\overline{B}(-L)\sin(\tilde{k}s + \chi_v L)$$
$$+\alpha\int\limits_{-L}^{s}\left\{\frac{\mathrm{i}\omega}{k\cos\psi}E_{0\tau}(s') + \overline{F}[s',\overline{A},\overline{B}]\right\}\sin\tilde{k}(s-s')\,\mathrm{d}s', \qquad (4.11)$$

where $\tilde{k} = k + \chi_v$. The four constants $\overline{A}(\pm L)$ and $\overline{B}(\pm L)$ in (4.11) may be found by the boundary conditions for the current $J(\pm L) = 0$ and by the symmetry conditions related to the method of vibrator excitation and its configuration.

Let us consider the symmetric case $E_{0\tau}(s) = E_{0\tau}(-s)$, $r(s) = r(-s)$. When the constants are substituted in (4.11), the current may be written as

$$J(s) = \alpha\frac{\mathrm{i}\omega}{k\cos\psi}\left\{\int\limits_{-L}^{s}E_{0\tau}(s')\sin\tilde{k}(s-s')\,\mathrm{d}s'\right.$$
$$\left. - \frac{\sin\tilde{k}(L+s) + \alpha P^\mathrm{s}[kr(s),\tilde{k}(L+s)]}{\sin 2\tilde{k}L + \alpha P^\mathrm{s}[kr\{s\},2\tilde{k}L]}\int\limits_{-L}^{L}E_{0\tau}(s')\,\sin\tilde{k}(L-s')\,\mathrm{d}s'\right\}, \qquad (4.12)$$
$$P^\mathrm{s}[kr(s),\tilde{k}(L+s)] = \int\limits_{-L}^{s}\left\{\frac{\mathrm{e}^{-\mathrm{i}k\tilde{R}[s',-L;r(s')]}}{\tilde{R}[s',-L;r(s')]} + \frac{\mathrm{e}^{-\mathrm{i}k\tilde{R}[s',L;r(s')]}}{\tilde{R}[s',L;r(s')]}\right\}\sin\tilde{k}(s-s')\,\mathrm{d}s'\Big|_{s=L}$$
$$= P^\mathrm{s}[kr\{s\},2\tilde{k}L].$$

Let a vibrator be excited by a plane electromagnetic wave with amplitude E_0: normally incident on it and $r(s) = r_0 + |s|\tan\psi$. Then finally,

$$J(s) = -\alpha\frac{\mathrm{i}\omega}{k\tilde{k}}E_0$$
$$\times\frac{(\cos\tilde{k}s - \cos\tilde{k}L) + \alpha\{\sin\tilde{k}LP^\mathrm{s}[kr(s),\tilde{k}(L+s)] - [1-\cos\tilde{k}(L+s)]P^\mathrm{s}[kr\{s\},\tilde{k}L]\}}{\cos\tilde{k}L + \alpha P^\mathrm{s}[kr\{s\},\tilde{k}L]}, \qquad (4.13)$$

$$P^\mathrm{s}[kr\{s\},\tilde{k}L] = \int\limits_{-L}^{L}\frac{\mathrm{e}^{-\mathrm{i}k\sqrt{(L-s)^2+r^2(s)}}}{\sqrt{(L-s)^2+r^2(s)}}\cos\tilde{k}s\,\mathrm{d}s.$$

The scattering properties of passive vibrators in free space are usually characterized by a normalized backscattering cross section (BSCS) σ/λ^2 ([11] in Chap. 2, [18] in Chap. 3, [13]), which may be written as

$$\frac{\sigma}{\lambda^2} = \frac{4\alpha^2}{\pi}\left|\frac{k}{\tilde{k}}\right|^4 \left|\frac{\sin \tilde{k}L}{\cos \tilde{k}L + \alpha P^{s}[kr\{s\}, \tilde{k}L]} - \tilde{k}L\right|^2 . \quad (4.14)$$

The vibrator current has both symmetric $J^{s}(s) = J^{s}(-s)$ and antisymmetric $J^{a}(s) = -J^{a}(-s)$ components, and thus $J(s) = J^{s}(s) + J^{a}(s)$ [see (2.13) in Chap. 2] for the oblique incidence of a plane wave or when $r(s) \neq r(-s)$.

Figure 4.2 shows a BSCS plot for a perfectly conducting ($\overline{Z}_{s} = 0$) biconical vibrator as a function of its electrical length $2L/\lambda$ ($2L$ =const $=$ 15 cm, $r_0 = 0.1$ cm) for different angles ψ. It can be seen that a passive scattering vibrator becomes more wideband when the angle ψ is increased. Note that the resonant $(2L/\lambda)_{res}$ values change slightly, but the maximal values of σ/λ^2 grow considerably even in the second peak. An analogous picture is observed for $\lambda =$ const, as shown in Fig. 4.3 for $\lambda = 10$ cm, $r_0 = 0.0127$ cm. Here the dotted curve and the circles mark the results obtained by the Hallen–King iterations method and the experimental data, respectively, for a silvered conductor from [13] (all curves are normalized by the maximal experimental BSCS value for the first resonance). Thus we can conclude that a biconical vibrator can effectively operate in a wider range of wavelengths than

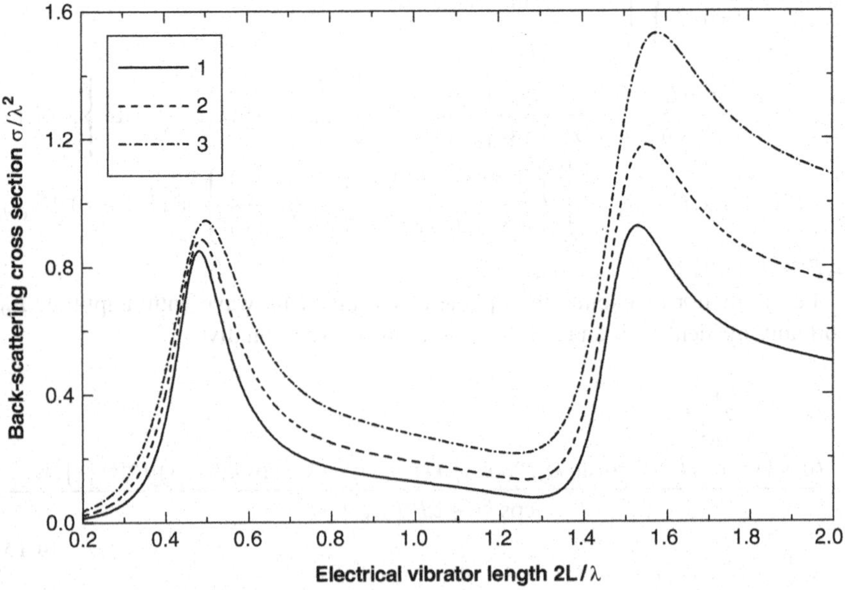

Fig. 4.2 Graph of σ/λ^2 versus the electric length of the vibrator for $2L$ =15 cm, $r_0 = 0.1$ cm and for different angles ψ: *1* $\psi = 0°$ ($r_L = 0.1$ cm); *2* $\psi = 1.1°$ (r_L =0.25 cm); *3* $\psi = 3.1°$ ($r_L = 0.5$ cm)

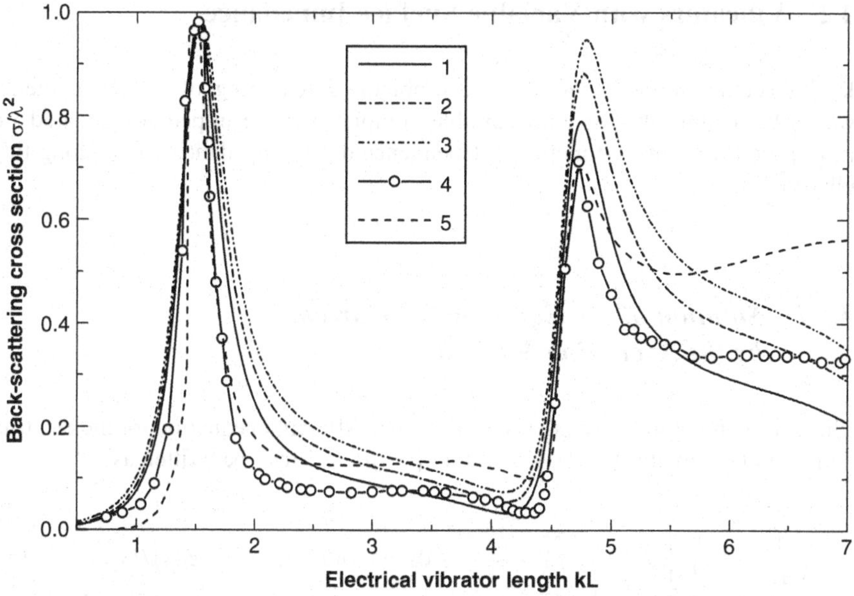

Fig. 4.3 Graph of σ/λ^2 versus kL for $\lambda = 10$ cm, $r_0 = 0.0127$ cm, and various r_L/r_0: 1 $r_L = r_0$; 2 $r_L = 5r_0$; 3 $r_L = 10r_0$; 4 and 5 the experimental data and the calculated values for $r_L = r_0$ [13]

Table 4.1 The calculated values and the experimental data of the maximal value of σ/λ^2 for thin metallic conductors

r (cm)	kr	Materials	z_i (ohm/cm)	σ/λ^2		
				Experiment ([13] in Chap. 2)	Calculation ([13] in Chap. 2)	Calculation (4.14)
003 × 1.27	0.0024	Copper	0.625 + i0.597	0.768	0.805	0.804
0.002 × 1.27	0.0016	Platinum	2.27 + i2.21	0.690	0.725	0.744
0.003 × 1.27	0.0024	Platinum	1.15 + i1.47	0.727	0.760	0.774
0.005 × 1.27	0.0040	Platinum	0.93 + i0.886	0.763	0.788	0.792

vibrators having a constant radius, while the resonant value $(2L/\lambda)_{\text{res}}$ is practically unchanged.

Naturally, the characteristics of an impedance vibrator made of metal with finite conductivity will change in comparison with the case of perfect conductivity. Table 4.1 represents the maximal value of BSCS calculated by (4.14) with the variation method ([13] in Chap. 2) and the experimental data ([13] in Chap. 2) for copper and platinum thin wires with different radii when condition of the first resonance is fulfilled. Obviously, the radius variation along the vibrators will give the same qualitative changes of the BSCS curves as shown in Figs. 4.2 and 4.3 for perfectly conducting biconical vibrators.

Note that the problem solution may be easily generalized to vibrators in an infinite material medium.

4.2 Vibrators with Variable Surface Impedance

In this section we will consider the problem of scattering of electromagnetic waves by a thin vibrator with variable complex surface impedance located in free space for arbitrary functional dependence of the impedance value along the vibrator.

4.2.1 Solution of the Equation for Current by the Averaging Method

The starting point for our analysis is the integrodifferential equation for the electric current in the vibrator (1.24), which for $\varepsilon_1 = \mu_1 = 1$ may be written as

$$\left(\frac{\mathrm{d}^2}{\mathrm{d}s^2}+k^2\right)\int\limits_{-L}^{L} J(s')\frac{\mathrm{e}^{-\mathrm{i}k\sqrt{(s-s')^2+r^2}}}{\sqrt{(s-s')^2+r^2}}\mathrm{d}s' = -\mathrm{i}\omega E_{0s}(s)+\mathrm{i}\omega z_i(s)J(s). \quad (4.15)$$

Let us solve the problem by the asymptotic averaging method, reducing (4.15) to an integrodifferential equation with small parameter [as in (2.2), (2.3)]. Thus we obtain a standard system of equations relative to $A(s)$ and $B(s)$, which after a change of variable may be written as (2.6):

$$\begin{aligned}\frac{\mathrm{d}A(s)}{\mathrm{d}s} &= -\frac{\alpha}{k}\{\mathrm{i}\omega E_{0s}(s)+F[s,A(s),B(s)]\\ &\quad -\mathrm{i}\omega z_i(s)[A(s)\cos ks+B(s)\sin ks]\}\sin ks,\\ \frac{\mathrm{d}B(s)}{\mathrm{d}s} &= +\frac{\alpha}{k}\{\mathrm{i}\omega E_{0s}(s)+F[s,A(s),B(s)]\\ &\quad -\mathrm{i}\omega z_i(s)[A(s)\cos ks+B(s)\sin ks]\}\cos ks.\end{aligned} \quad (4.16)$$

Carrying out partial averaging in (4.16), we obtain the averaged equations of the first approximation in the small parameter α:

$$\begin{aligned}\frac{\mathrm{d}\bar{A}(s)}{\mathrm{d}s} &= -\alpha\left\{\frac{\mathrm{i}\omega}{k}E_{0s}(s)+\bar{F}[s,\bar{A},\bar{B}]\right\}\sin ks+\chi_a\bar{A}(s)+\chi_s\bar{B}(s),\\ \frac{\mathrm{d}\bar{B}(s)}{\mathrm{d}s} &= +\alpha\left\{\frac{\mathrm{i}\omega}{k}E_{0s}(s)+\bar{F}[s,\bar{A},\bar{B}]\right\}\cos ks-\chi_s\bar{A}(s)-\chi_a\bar{B}(s).\end{aligned} \quad (4.17)$$

Here we have considered that the impedance has symmetric $(z_i^{\mathrm{s}}(s))$ and antisymmetric $(z_i^{\mathrm{a}}(s))$ $z_i(s) = z_i^{\mathrm{s}}(s)+z_i^{\mathrm{a}}(s)$ components, which were used to define

$$\chi_s = \lim_{2L\to\infty} \frac{1}{2L}\frac{\alpha i\omega}{k}\int_{-L}^{L} z_i^{s}(s)\sin^2 ks\{\cos^2 ks\}\,ds,$$

$$\chi_a = \lim_{2L\to\infty} \frac{1}{2L}\frac{\alpha i\omega}{k}\int_{-L}^{L} \frac{1}{2} z_i^{a}(s)\sin 2ks\,ds.$$

The solution of the system of (4.17) may be obtained in the form ([1] in Chap. 2).

$$\begin{aligned}\overline{A}(s) &= \chi_s[C_1(s)\sin\chi s + C_2(s)\cos\chi s],\\ \overline{B}(s) &= -\chi_a[C_1(s)\sin\chi s + C_2(s)\cos\chi s] + \chi[C_1(s)\cos\chi s - C_2(s)\sin\cos\chi s].\end{aligned} \tag{4.18}$$

Here $\chi^2 = \chi_s^2 - \chi_a^2$, and $C_1(s)$ and $C_2(s)$ are new unknown functions. We obtain the functions $C_1(s)$ and $C_2(s)$, and then $\overline{A}(s)$ and $\overline{B}(s)$, after substituting (4.18) into the system (4.17):

$$\begin{aligned}\overline{A}(s) =& \frac{1}{\chi}\Bigg\{\overline{A}(-L)[\chi\cos\chi(L+s)+\chi_a\sin\chi(L+s)]+\overline{B}(-L)[\chi_s\sin\chi(L+s)]\\ &+\alpha\int_{-L}^{s}\left\{\frac{i\omega}{k}E_{0s}(s')+\overline{F}[s',\overline{A},\overline{B}]\right\}\\ &\times[\chi_s\cos ks'\sin\chi(s-s')-\chi\sin ks'\cos\chi(s-s')+\chi_a\sin ks'\sin\chi(s-s')]\,ds'\Bigg\},\\ \overline{B}(s) =& \frac{1}{\chi}\Bigg\{-\overline{A}(-L)[\chi_s\sin\chi(L+s)]+\overline{B}(-L)[\chi\cos\chi(L+s)-\chi_a\sin\chi(L+s)]\\ &+\alpha\int_{-L}^{s}\left\{\frac{i\omega}{k}E_{0s}(s')+\overline{F}[s',\overline{A},\overline{B}]\right\}[\chi_s\sin ks'\sin\chi(s-s')\\ &+\chi\cos ks'\cos\chi(s-s')-\chi_a\cos ks'\sin\chi(s-s')]\,ds'\Bigg\},\end{aligned} \tag{4.19}$$

where

$$\begin{aligned}\overline{F}[s,\overline{A},\overline{B}] =& [\overline{A}(L)\sin kL - \overline{B}(L)\cos kL]\frac{e^{-ik\sqrt{(L-s)^2+r^2}}}{\sqrt{(L-s)^2+r^2}}\\ &+[\overline{A}(-L)\sin kL + \overline{B}(-L)\cos kL]\frac{e^{-ik\sqrt{(L+s)^2+r^2}}}{\sqrt{(L+s)^2+r^2}}.\end{aligned}$$

Substituting $\overline{A}(s)$ and $\overline{B}(s)$ as the approximating functions for $A(s)$ and $B(s)$ into the formula $J(s) = A(s)\cos ks + B(s)\sin ks$, we obtain a general asymptotic (in the parameter α) expression for the current in a thin vibrator with variable impedance along its axis and arbitrary excitation

$$\begin{aligned} J(s) = {} & \overline{A}(-L)[\cos ks \cos\chi(L+s) - (\overline{\chi}_s \sin ks - \overline{\chi}_a \cos ks)\sin\chi(L+s)] \\ & + \overline{B}(-L)[\sin ks \cos\chi(L+s) + (\overline{\chi}_s \cos ks - \overline{\chi}_a \sin ks)\sin\chi(L+s)] \\ & + \alpha \int\limits_{-L}^{s} \left\{ \frac{i\omega}{k} E^{s}_{0s}(s') + \overline{F}[s', \overline{A}(\pm L), \overline{B}(\pm L)] \right\} \\ & \times \{\sin k(s-s')\cos\chi(s-s') \\ & + [\overline{\chi}_s \cos k(s-s') - \overline{\chi}_a \sin k(s+s')]\sin\chi(s-s')\} ds'. \end{aligned} \tag{4.20}$$

Here $\overline{\chi}_s = \chi_s/\chi$, $\overline{\chi}_a = \chi_a/\chi$. The four constants $\overline{A}(\pm L)$ and $\overline{B}(\pm L)$ are defined by boundary and symmetry conditions directly related to the method of vibrator excitation and by the impedance variation along the vibrator. Note that the expressions for the current (4.20) are rather bulky for practical use, except for the case of constant impedance along the vibrator when $\chi = \alpha(i\omega/2k)z_i$, $\overline{\chi}_s = 1$, $\overline{\chi}_a = 0$, and (4.20) is sufficiently simplified. Bearing in mind these remarks, we obtain the solution of (4.15) by the induced EMF method [14] in the next section.

4.2.2 Solution of the Equation for Current by the Induced EMF Method

Let us apply the induced EMF method for an approximate analytical solution of (4.15) for symmetric vibrator excitation $E_{0s}(s) = E_{0s}(-s)$ and under the condition $z_i(s) = z_i(-s)$. Let us approximate the electric current distribution by the expression

$$J(s) = J_0 f(s), \quad f(\pm L) = 0, \tag{4.21}$$

where J_0 is the unknown current amplitude and $f(s)$ is the given function. Then multiplying the left and the right sides of (4.15) by $f(s)$, we obtain, after integration,

$$J_0 = \frac{-\frac{i\omega}{2k}\int\limits_{-L}^{L} f(s)E_{0s}(s)\,ds}{\frac{1}{2k}\int\limits_{-L}^{L} f(s)\left[\left(\frac{d^2}{ds^2}+k^2\right)\int\limits_{-L}^{L} f(s')\frac{e^{-ikR(s,s')}}{R(s,s')}ds'\right]ds - \frac{i\omega}{2k}\int\limits_{-L}^{L} f^2(s)z_i(s)\,ds},$$

$$R(s,s') = \sqrt{(s-s')^2 + r^2}. \tag{4.22}$$

As is well known, the induced EMF method ([11] in Chap. 1, [14]) gives a more precise solution of the integral equation, the better the approximating functions for the current. We suspect that a functional dependence of surface impedance must include information about mean impedance and will best correspond to a real physical process (it is obvious for small oscillations of the impedance value near the mean value). Variable impedance influences the current amplitude J_0 and hence other vibrator characteristics. As will be shown below, two different distributions of impedance with the same mean value give different values of the backscattering cross section.

Let us consider the normal incidence of a plane electromagnetic wave $E_{0s}(s) = E_0$ on a vibrator. Then it is possible to choose the current distribution function $f(s)$ (2.30) obtained by the averaging method for a receiving vibrator with constant surface impedance in free space as

$$f(s) = \cos \tilde{k}s - \cos \tilde{k}L, \tag{4.23}$$

where $\tilde{k} = k - (\mathrm{i}2\pi z_i^{\mathrm{av}}/Z_0\Omega)$, $z_i^{\mathrm{av}} = (1/2L)\int\limits_{-L}^{L} z_i(s)\,\mathrm{d}s$ is the mean value of the internal impedance, and $\Omega = 2\ln(2L/r)$. Note that (4.23) contains direct information about the impedance distribution $z_i(s)$, while the current distribution approximating functions used by other investigators [7–12] do not. The verification of the choice of such a function $f(s)$ is presented in Sect. 4.2.3, where the distribution functions for $E_{0s}(s) = E_{0s}(-s)$ and $z_i(s) \neq z_i(-s)$ are given.

After substituting (4.23) into (4.22) and calculating J_0, we obtain an expression for the current:

$$J(s) = -\frac{\mathrm{i}\omega}{k\tilde{k}} E_0 \frac{(\sin \tilde{k}L - \tilde{k}L \cos \tilde{k}L)(\cos \tilde{k}s - \cos \tilde{k}L)}{Z(kr, \tilde{k}L) + F_z(\tilde{k}r, \tilde{k}L)}, \tag{4.24}$$

where

$$Z(kr, \tilde{k}L) = \frac{1}{2k} \int\limits_{-L}^{L} (\cos \tilde{k}s - \cos \tilde{k}L)\left(\frac{\mathrm{d}^2}{\mathrm{d}s^2} + k^2\right) F_{\mathrm{f}}(s)\mathrm{d}s, \tag{4.25}$$

$$F_{\mathrm{f}}(s) = \int\limits_{-L}^{L} (\cos \tilde{k}s' - \cos \tilde{k}L)\frac{\mathrm{e}^{-\mathrm{i}kR(s,s')}}{R(s,s')}\mathrm{d}s', \tag{4.26}$$

$$F_z(\tilde{k}r, \tilde{k}L) = -\frac{\mathrm{i}}{r} \int\limits_{-L}^{L} (\cos \tilde{k}s - \cos \tilde{k}L)^2 \overline{Z}_S(s)\mathrm{d}s. \tag{4.27}$$

Here $\overline{Z}_S(s) = \overline{R}_S(s) + i\overline{X}_S(s) = (2\pi r z_i(s))/Z_0$ is the normalized surface impedance of the vibrator distributed over the vibrator as $\overline{Z}_S(s) = \overline{Z}_S\phi(s)$, where $\phi(s)$ is the given function. In particular, $\overline{Z}_S^{av} = \overline{Z}_S$ when $\phi(s) = 1$, and

$$F_z(\tilde{k}r, \tilde{k}L) = -\frac{i\overline{Z}_S}{\tilde{k}r}\left[\tilde{k}L(2 + \cos 2\tilde{k}L) - \frac{3}{2}\sin 2\tilde{k}L\right]. \qquad (4.28)$$

Let us obtain an expression for $Z(kr, \tilde{k}L)$. Integrating (4.25) by parts in:

$$Z(kr, \tilde{k}L) = \frac{1}{2k}\left\{-\tilde{k}\sin\tilde{k}sF_f(s)\Big|_{-L}^{L} + \int\limits_{-L}^{L}[(k^2 - \tilde{k}^2)\cos\tilde{k}s - k^2\cos\tilde{k}L]F_f(s)\,ds\right\}.$$

The function $F_f(s)$can be obtained from (5.31) and (5.32) in [15] when the inequality $kr \ll 1$ holds:

$$\begin{aligned}F_f(s) = {} & \frac{1}{2}\cos\tilde{k}s\{2\Omega(s) - \operatorname{Cin}k^+(L+s) - \operatorname{Cin}k^+(L-s) - \operatorname{Cin}k^-(L+s)\\ & - \operatorname{Cin}k^-(L-s) - i[\operatorname{Si}k^+(L+s) + \operatorname{Si}k^+(L-s) + \operatorname{Si}k^-(L+s)\\ & + \operatorname{Si}k^-(L-s)]\} + \frac{1}{2}\sin\tilde{k}s\{\operatorname{Si}k^+(L+s) - \operatorname{Si}k^+(L-s)\\ & - \operatorname{Si}k^-(L+s) + \operatorname{Si}k^-(L-s) - i[\operatorname{Cin}k^+(L+s) - \operatorname{Cin}k^+(L-s)\\ & - \operatorname{Cin}k^-(L+s) + \operatorname{Cin}k^-(L-s)]\} - \cos\tilde{k}L\{\Omega(s) - \operatorname{Cin}k(L+s)\\ & - \operatorname{Cin}k(L-s) - i[\operatorname{Si}k(L+s) + \operatorname{Si}k(L-s)]\}.\end{aligned} \qquad (4.29)$$

Here $k^+ = k + \tilde{k}$, $k^- = k - \tilde{k}$, $\Omega(s) = \ln\left[\frac{\left(\sqrt{(L+s)^2 + r^2} + (L+s)\right)}{\left(\sqrt{(L-s)^2 + r^2} - (L-s)\right)}\right]$.
Then finally,

$$Z(kr, \tilde{k}L) = \left(\frac{\tilde{k}}{k}\right)\sin\tilde{k}LF_f(L) - \frac{k}{2}\cos\tilde{k}L\int\limits_{-L}^{L}F_f(s)\,ds + \frac{k^2 - \tilde{k}^2}{2k}\int\limits_{-L}^{L}\cos\tilde{k}s\,F_f(s)ds, \qquad (4.30)$$

which for $\tilde{k} = k$ may be reduced to the familiar expression from ([11] in Chap. 2).

The expression for the backscattering cross section σ/λ^2 is given by

$$\frac{\sigma}{\lambda^2} = \frac{4}{\pi}\left|\frac{k}{\tilde{k}}\right|^4 \left|\frac{\left(\sin\tilde{k}L - \tilde{k}L\cos\tilde{k}L\right)^2}{Z(kr,\tilde{k}L) + F_z(\tilde{k}r,\tilde{k}L)}\right|^2. \tag{4.31}$$

Let us consider two functions with equal means to illustrate the influence of the impedance variation along the vibrator on its characteristics:

1. $\phi_1(s) = e^{-\beta\frac{|s|}{L}}$, the distribution decreasing to the vibrator ends exponentially.
2. $\phi_2(s) = e^{\beta\left(\frac{|s|}{L}-1\right)}$, the distribution increasing to the vibrator ends exponentially.

Here β is an arbitrary dimensionless constant,

$$\overline{\phi_{1,2}(s)} = \frac{1}{2L}\int_{-L}^{L}\phi_{1,2}(s)\,ds = \frac{1-e^{-\beta}}{\beta}, \tag{4.32}$$

and the plots of the functions $\phi_{1,2}(s)$and their mean values for different values of β are given in Fig. 4.4.

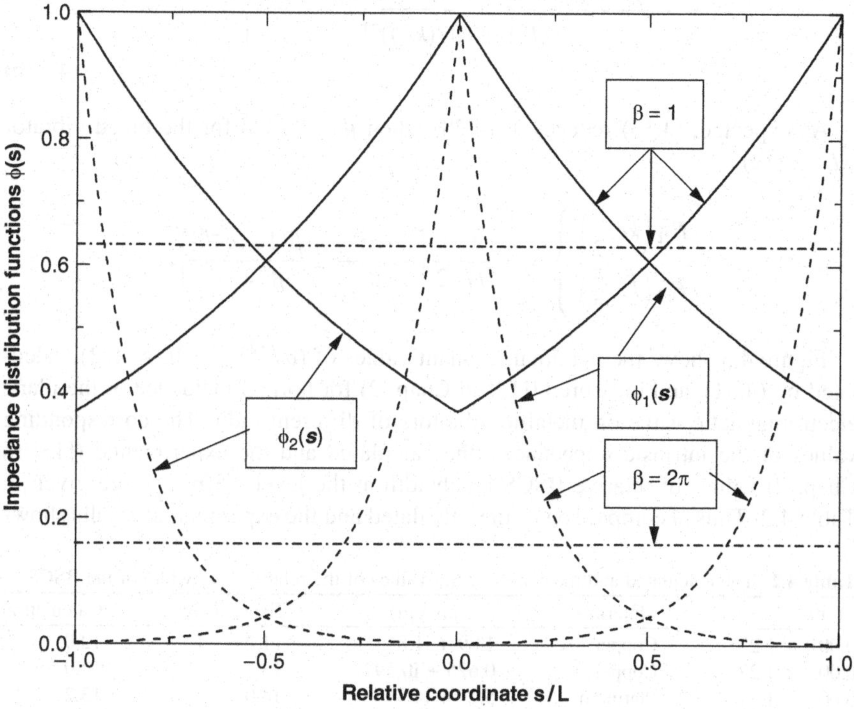

Fig. 4.4 The impedance variation along the vibrator: *the solid curves* correspond to $\beta = 1$; *the dotted curves* correspond to $\beta = 2\pi$; *the dash-dotted lines* represent the mean values

Then $\overline{Z}_{S_1}^{av} = \overline{Z}_{S_2}^{av} = \overline{Z}_S\overline{\phi_{1,2}(s)}$ and

$$F_{z1}(\tilde{k}r, \tilde{k}L) = -\frac{i\overline{Z}_S}{\tilde{k}r} \times \left\{ \frac{1-e^{-\beta}}{\beta}\tilde{k}L(2+\cos 2\tilde{k}L) + \frac{e^{-\beta}\left(\tilde{k}L\sin 2\tilde{k}L - \frac{\beta}{2}\cos 2\tilde{k}L\right) + \frac{\beta}{2}}{2\tilde{k}L[1+(\beta/(2\tilde{k}L))^2]} - \frac{4\cos\tilde{k}L\left[e^{-\beta}(\tilde{k}L\sin\tilde{k}L - \beta\cos\tilde{k}L) + \beta\right]}{\tilde{k}L[1+(\beta/(\tilde{k}L))^2]} \right\}, \tag{4.33a}$$

$$F_{z2}(\tilde{k}r, \tilde{k}L) = -\frac{i\overline{Z}_S}{\tilde{k}r} \times \left\{ \frac{1-e^{-\beta}}{\beta}\tilde{k}L(2+\cos 2\tilde{k}L) + \frac{\tilde{k}L\sin 2\tilde{k}L + \frac{\beta}{2}(\cos 2\tilde{k}L - e^{-\beta})}{2\tilde{k}L[1+(\beta/(2\tilde{k}L))^2]} - \frac{4\cos\tilde{k}L\left[\tilde{k}L\sin\tilde{k}L + \beta(\cos\tilde{k}L - e^{-\beta})\right]}{\tilde{k}L[1+(\beta/(\tilde{k}L))^2]} \right\}. \tag{4.33b}$$

As expected, (4.33) reduces to (4.28) when $\beta = 0$, and for the tuned vibrator ($\tilde{k}L = \pi/2$),

$$\left.\begin{array}{l} F_{z1}\left(\tilde{k}r, \frac{\pi}{2}\right) \\ F_{z2}\left(\tilde{k}r, \frac{\pi}{2}\right) \end{array}\right\} = -\frac{i\overline{Z}_S}{\tilde{k}r}\frac{\pi}{2}\left[\frac{1-e^{-\beta}}{\beta} \pm \frac{\beta(1+e^{-\beta})}{\beta^2+\pi^2}\right].$$

Figure 4.5 shows the maximal resonant values of $(\sigma/\lambda^2)_{\max}$ ($2L \approx \lambda/2$), calculated by (4.31), and measured ([13] in Chap. 2) for normal incidence of the plane electromagnetic wave on metallic vibrators of different radii. The corresponding values of the intrinsic impedances, the calculated and the experimental ([13] in Chap. 2) values of relative BSCS bandwidth at the level $0.5(\sigma/\lambda^2)$, are given in Table 4.2. Thus, a comparison of the calculated and the experimental results shows

Table 4.2 The calculated and the experimental values of the relative bandwidth of the BSCS

r (cm)	Materials	z_i (ohm/cm)	Experiment	Calculation
0.001 × 1.27	Copper	1.89 + i1.8	11.2	12.0
0.003 × 1.27	Copper	0.625 + i0.597	13.5	14.0
0.001 × 1.27	Platinum	5.27 + i4.58	14.0	13.2
0.002 × 1.27	Platinum	2.27 + i2.21	14.4	13.7
0.003 × 1.27	Platinum	1.15 + i1.47	15.3	14.1
0.005 × 1.27	Platinum	0.93 + i0.886	16.4	15.5
0.001 × 1.27	Bismuth	29.4 + i7.0	24.7	24.8

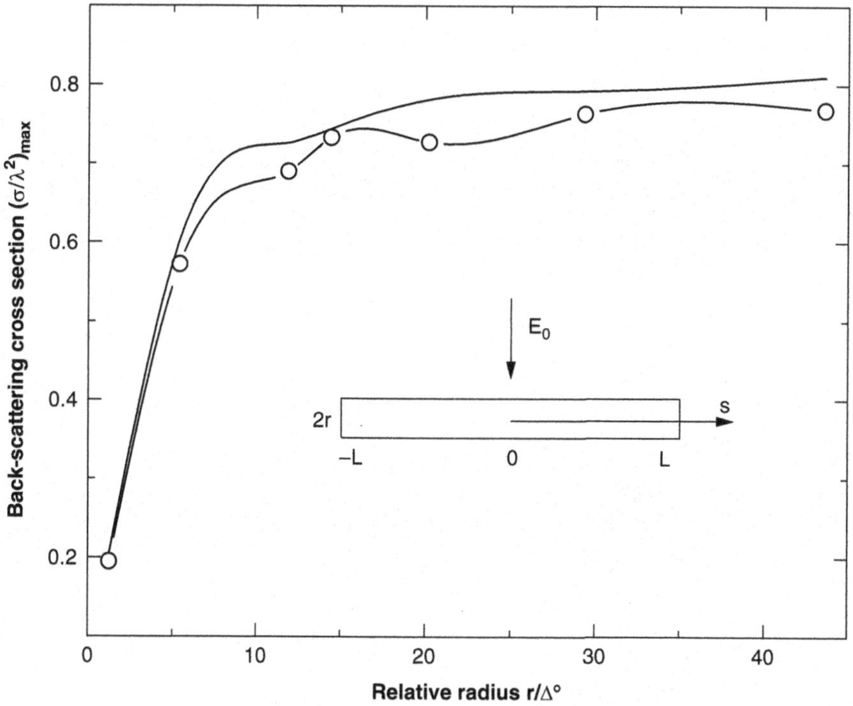

Fig. 4.5 $BSCS_{max}$ for metallic vibrators with different radii (see Table 4.2) for the normal incidence of a plane electromagnetic wave with amplitude E_0 on a vibrator of radius r and length $2L$: *the solid curve* is for the calculation, *the circles* for the experimental data ([13] in Chap. 2) (Δ° is the skin-layer thickness)

that the mathematical model and the problem solution by the induced EMF method completely correspond to real physical processes.

The BSCS values as a function of the electrical length for perfectly conducting (curve 1) and impedance (curves 2–5) vibrators with constant impedance distribution along the vibrator are presented in Fig. 4.6. The experimental values from [13] (the silvered wire, $r = 0.1016$ cm, $\lambda = 10$ cm) are also given here, marked by circles. Note that all curves here and below in this section are normalized by the maximal (resonant) experimental value of σ/λ^2. As expected, the BSCS value decreases when the real part of surface impedance $\overline{R}_S$ increases (curves 2 and 3) in comparison to a perfectly conducting vibrator, but the resonant length $2L_{res}$ remains practically unchanged. The inductive impedance (curve 4) decreases $2L_{res}$, while capacitive impedance (curve 5) increases $2L_{res}$ and the relative bandwidth of the BSCS.

The influence of irregularity in the impedance for several values of inductive and capacitive surface impedance and for different values of the parameter β ($\overline{R}_S^{av} = 0.005$, $\overline{X}_S^{av} = \pm 0.05$, $\beta = 1;\ 2\pi$) may be observed in Fig. 4.7. As can be seen, the difference between the values of σ/λ^2 for the constant distribution increases when β is increased, and the distribution decreasing to the vibrator's